CROCKETT

CROCKETT
A Bio-Bibliography

Richard Boyd Hauck

Popular Culture Bio-Bibliographies

Greenwood Press
Westport, Connecticut • London, England

Copyright Acknowledgment

The author would like to thank the University of Tennessee Press for the use of scattered quotes from the facsimile pages of *A Narrative of the Life of David Crockett of the State of Tennessee* by David Crockett with annotations and introduction by James A. Shackford and Stanley J. Folmsbee, copyright © 1973 by The University of Tennessee Press.

Library of Congress Cataloging in Publication Data

Hauck, Richard Boyd, 1936-
   Crockett, a bio-bibliography.

   (Popular culture bio-bibliographies, ISSN 0193-6891)
   Bibliography: p.
   Includes index.
   1. Crockett, Davy, 1786-1836—Bibliography.
2. Pioneers—United States—Biography. 3. Crockett, Davy, 1786-1836—Legends—Bibliography. I. Title.
II. Series.
Z8199.63.H38   [F436.C95]   016.9768'04'0924 [B]   82-6109
ISBN 0-313-22272-X (lib. bdg.)   AACR2

Copyright © 1982 by Richard Boyd Hauck

All rights reserved. No portion of this book may be reproduced, by any process or technique, without the express written consent of the publisher.

Library of Congress Catalog Card Number: 82-6109
ISBN: 0-313-22272-X
ISSN: 0193-6891

First published in 1982

Greenwood Press
A division of Congressional Information Service, Inc.
88 Post Road West, Westport, Connecticut 06881

Printed in the United States of America

10 9 8 7 6 5 4 3 2 1

For Sarah—

Who can whip her weight in wildcats but would rather make them heroes in her own tall tale.

CONTENTS

Plates	ix
Preface	xi
Acknowledgments	xiii
Note on Citations	xv
Introduction: A Common Man, an Uncommon Style	xvii

1 David Crockett—The Facts   3

The Art of Autobiography	4
A Birthright of Poverty	9
Quick Wits and Fast Feet	11
Getting "Made up" in Life	14
Indians!	18
The Labyrinth	26
Starting Over Again, and Again	28
Law and Order—Crockett Style	31
Bears!	33
In the Public Interest	35
Neck or Nothing	39
Crockett vs. Washington	40
Meeting Himself Coming Back	44
A World of Country	47
The Many Deaths of Crockett	50

2 Davy Crockett—The Fictions — 55

The Style of the Man — 60
Half Horse, Half Alligator — 67
The Almanacs — 79
Tall Tales — 83
Actors Acting the Actor — 88
Truthful Fictions — 96
The Continuing Invention of Crockett — 104

3 The Crockett Idiom — 114

George Russell — 116
The Tall Tale Reversed — 117
Bears, Wrecks, and Chaos — 119
Two Useful Coonskins — 125
With the Bark Off — 129
Crockett's First Speech in Congress — 130
The Dog and the Raccoon — 131

4 The Crockett Record — 134

5 The Crockett Chronology — 146

Bibliography — 151
Index — 159

PLATES

1. Oil portrait of Congressman David Crockett in hunting garb, by John Gadsby Chapman, painted in 1834 — 107
2. John Gadsby Chapman's romanticized bust portrait of David Crockett, painted in 1834, in preparation for the preceding full-length portrait — 108
3. A contrasting view of David Crockett. Engraving by Chiles and Lehmann, Lithographers, Philadelphia, 1834, from an oil painting by Samuel Stillman Osgood — 109
4. James Hackett as Nimrod Wildfire in *The Lion of the West*. Engraving made from a portrait by Ambrose Andrews. Captioned: "Come back, stranger! or I'll plug you like a watermillion!" — 110
5. Woodcut portrait of Davy apparently copied from Andrews's portrait of Hackett. Cover of *Davy Crockett's Almanack*, 1837 — 111
6. Fess Parker as Davy Crockett and Buddy Ebsen as Georgie Russel in *Davy Crockett and the River Pirates*, 1956 — 112
7. John Wayne as David Crockett and Laurence Harvey as William Travis in *The Alamo*, 1960 — 113

PREFACE

As every student interested in the whole story has discovered, the history of the legendary Davy Crockett is a more complex subject than the history of the actual David Crockett. Scholars committed to finding the raw facts have been frustrated and even angered by the density of the legendary material that stands between them and the historical man. On the other hand, literary historians, students of popular culture and American humor, folklorists, and historical novelists have invariably found the factual life of Crockett to be a rather dull subject, and they have gladly written about a different kind of reality, the living legend itself. Paradoxically, serious historians would probably not bother to pursue the real Crockett at all if it were not for the tremendous popularity and historical importance of the legend. The man lived one life, while the legend has always had a vigorous life of its own. The legend did begin with actual qualities of Crockett's character and actual events, but it achieved its independent life even before he died, and it has since taken shapes that often contradict and resist the facts of his life and death. In one sense, then, the two lives are quite distinct: the life of the man was bound by circumstance and time; the life of the legend seems bound only by the dynamic imagination of the audience that keeps the legend alive.

The two lives are, however, strongly tied together by certain abstractions. The craziness of frontier life is reflected both in the actual life of David Crockett and in the wacky humor that permeates the stories of Davy Crockett. The heroism in both the life and the legend, especially as it is exemplified by the many and various stories of the Alamo, provides an important but less substantial link. The most tangible connection between Crockett the man and Crockett the legend lies with style. The rough, satirical, irreverent, iconoclastic, and ever-optimistic streak of frontier humor displayed in the man's yarn spinning, electioneering, and writing defines exactly the spirit of his legend. It also reflects the spirit of the milieu from which he came and the genre of American humor to which he contributed. This humor has much to do with place and time. Settlers cultivated humorous language as a creative reaction to the chaos and hardship of life along the frontiers. Backwoods storytellers saw their art as the antithesis of all things formal, traditional, and authoritative, and the targets at which their humor was aimed included serious literature, proper grammar, and civilized manners. The American West in which this humor flourished was the scene out of which Crockett the man emerged; the legend of Crockett is in turn embedded in the myth of the American West.

For purposes of understanding both the life of the man and the life of the legend, this volume necessarily contains two biographies, one chapter for each life. But as the first section of chapter 2 indicates, the whole story of Crockett is told only when the two are set in a proper relation to one another. David Crockett lived and died, and his story is closed. Davy Crockett was born out of that life, and his story remains open. This volume is strenuously devoted to both the truth of the man's real life and the truth of the legend's real life, but this book also must inevitably become a contribution to the truthful fiction that is the unbounded and unclosed life of Crockett.

ACKNOWLEDGMENTS

I wish first to thank my wife, Dean Margaret Hauck, who always backs up her loving encouragement with concrete contributions to my research and editing. My gratitude and admiration go to our daughters, Margaret and Sarah, who not only helped with manuscript preparation, but always understood. Others who gave me invaluable assistance are Michael A. Lofaro, University of Tennessee; John T. Flanagan, George Hendrick, and Robert W. Rogers, University of Illinois; David Crockett, Elizabethton, Tennessee; Bob Cochran, University of Arkansas; Marc Reisch, Stony Brook, New York; Mary Ann Wimsatt, Southwest Texas State University; Charles Long, curator of the Alamo Museum; Fred Nash, rare book librarian, and Mary Ceibert and Louise Fitton, at the rare book room of the University of Illinois Library; David R. Smith, archivist, Walt Disney Archives; Casandra LaSalandra, Humanities Research Center, University of Texas; the staff of the Lilly Library, Indiana University; Bodil Gilliam, David Doerrer, Jean Kamerman, and many other patient colleagues on the staff of the University of West Florida Library. Assistance for research was provided by Lucius Ellsworth, dean of the College of Arts and Sciences, University of West Florida, and his office, which manages support provided in connection with the Abe Levin Professorship in the Humanities. Indispensable and continuous help came from the faculty and staff of the University of West Florida's English department and its chairman, Stan Millet, and from all my students, who wanted to know if Crockett was real and thus had opportunities to respond to my lectures on two kinds of reality. Special thanks to Professor M. Thomas Inge, Clemson University, whose foresight and wisdom made it possible for this book to "Go Ahead."

NOTE ON CITATIONS

In lieu of footnotes, I have used parenthetical citations which refer to sources by the last names of the authors, or by coded initials for the two books indicated below. I have added short titles where there would otherwise be ambiguities. Information for every citation is in the bibliography at the end of the book.

One of my intentions has been to offer a new interpretation of the facts of Crockett's life. As a result, I have made considerable use of Crockett's autobiography and of Shackford's definitive modern biography; and all citations of these two books are coded *N* and *DC*, respectively:

- *N* = David Crockett. *A Narrative of the Life of David Crockett of the State of Tennessee.* Edited by James A. Shackford and Stanley J. Folmsbee. Knoxville: University of Tennessee Press, 1973. First published in 1834.

- *DC* = James A. Shackford. *David Crockett: The Man and the Legend.* Chapel Hill: University of North Carolina Press, 1956.

Two of the Crockett books published in the 1830s are widely available only on microfilm or in later editions and reprints with various pagination, so I have chosen to cite chapter numbers rather than page numbers for Mathew St. Clair Clarke's *Sketches and Eccentricities of Colonel David Crockett of West Tennessee* (1833) and for Richard Penn Smith's *Col. Crockett's Exploits and Adventures in Texas* (1836).

INTRODUCTION: A COMMON MAN, AN UNCOMMON STYLE

To his great audience, the innumerable inheritors of his legend, David Crockett's true presence lives in an indefinite, fluid story that has been expanded and improved rather than diminished by time. Only recently have serious historians set aside the many permutations of the fable so that the long-obscured facts about the man's life might be discovered and made public. For instance, any person who has read the various fictions about the legendary "Davy" or seen him portrayed in movies or on television is likely to believe that in 1830 Crockett rose in the United States House of Representatives, almost alone against the entire body and absolutely alone against his Tennessee colleagues, to give a dramatic and impassioned speech opposing President Andrew Jackson's infamous Indian removal bill, which called for an appropriation of half a million dollars for moving Indians living in the Southeast to the desolate western areas of the Louisiana Purchase. But historians know that while Crockett did write such a speech, or had it written for him, this fiery declaration of human rights is most conspicuously absent from *The Register of Debates in Congress*. However, a June 1830 issue of the West Tennessee newspaper *Jackson Gazette* printed a summary of the speech and reported that Crockett had given it on May 24, the day the bill was passed, and a collection of congressional speeches on the Indian removal bill published in Boston later in 1830 included the same summary, giving the date of Crockett's oration as May 19. Apparently Crockett did give the speech—his tone would have been cantankerous rather than heroic—and then made some arrangement to keep it out of the official record. (The summary is reprinted in Blair, *Davy Crockett*; see also *DC*, pp. 116-26; and *N*, p. xv.)

These banal details are scarcely the stuff of legend. Could it be that the record of Crockett's life does not support the record of his legend? The story of what seems actually to have happened during the Indian removal controversy tells us that the historical Crockett was simply a good man caught up in the mundane machinations of politics. Perplexed by the deceptive phrasing and veiled purposes of the bill, and reacting with his usual hot impulsiveness against the Jacksonites' suspicious maneuvers, Crockett wrote the speech, with some help, and gave it in support of his negative vote. He warned that the bill was both morally wrong and grossly impractical: trying to move a whole population would bring disaster, not only to the Indians, but to the government—which would have to execute the plan and pay for it. Later, having second thoughts about what his defense of the Indians and opposition to Jackson implied for his political future, he persuaded the editors of the *Register of Debates*, two men whom he thought of as allies, to keep the speech out of print. The editors, Joseph Gale and William Seaton, did what he asked, but they may well have been the persons who leaked the report to the *Jackson Gazette* (*N*, p. xv and fn. 19).

Crockett was understandably reluctant to let the pioneer farmers among his constituents know that he had argued and voted against a bill which many of them favored wholeheartedly. In general, they believed that moving the Indians out would improve their chances to win the continuing battle for ownership of the new lands in Tennessee. Crockett realized he would not be able to convince those constituents who were pro-Jackson and anti-Indian that his position was both moral and practical. The apprehension that his stand would cost him votes in the next election was correct: he was not returned to Congress in 1831. He made a comeback in 1833, but was defeated a second time, in 1835, again largely because of his obstinate, vociferous, and sadly unproductive opposition to Jackson.

Worst of all, Crockett lost the congressional argument as well as the next election. The forced relocation of the Indians became history, commencing during the 1830s in that awesome atrocity the Indians named the Trail of Tears, and disintegrating during the 1840s in the Seminole War, a costly debacle for the United States Army. Although he was entirely unsuccessful in opposing Indian removal, Crockett's prediction that the move would be a disaster proved correct. The net result was upheaval, anguish, and death for the Indians, and failure, defeat, and humiliating losses for the soldiers sent to implement the outrageous law.

The unromantic factual biography of Crockett is riddled with such disappointments and stark realities, but its unheroic details are fascinating in their own right, since they show so clearly that life on the frontier was often nasty and ridiculous rather than glorious, and because, taken together, the details reveal how a tough, honest, and intelligent backwoodsman could be shredded in the machinery of politics and enmeshed in the absurd complexities of history. Crockett's failure to stop or even slow the progress of the Indian removal bill is emblematic of his whole public career.

Although it is certainly no epic, Crockett's story might well begin *in medias res* with this incident of failure, because it vividly illustrates how the facts behind a legend seldom meet the demands of the human imagination. At the same time, the facts of his life provide abundant evidence of the man's moral fiber. The biography of Crockett is not the history of an institution builder, conqueror, king, or president. Instead, it is a story of a common man who fought with uncommon style. In his role as a lone dissenter taking large personal risks, Crockett displayed the qualities Americans identify as those values which distinguish the type of the frontier individualist. These values are the seeds of his legend, and it is Crockett's legend which is the epic that has engaged his audience.

The Indians who were to be removed are not the imaginary wild Indians of nineteenth-century dime novels and twentieth-century Hollywood films. They constituted five tribes, more properly called nations, whose forebears had lived for many thousands of years in the southeastern area of what is now the United States. They were the Cherokees, Chickasaws, Choctaws, Creeks, and Seminoles—ancient civilizations rich with tradition, lore, and law. For a number of generations, they had endured a drawn-out, vicious era of war and broken treaties, and had found ways to accommodate themselves to the new patterns of living imposed upon them by the steady intrusions of white civilization. Most of the Indians living in or near the older settlements were indistinguishable from their pioneer neighbors in their style of living and, surprisingly often, in their physical appearance, since intermarriages were fairly common. In several towns and districts of Tennessee, Alabama, Georgia, and North Florida, Indians owned and farmed the land, built log cabins or rectangular houses of sawed lumber, obeyed the law, and otherwise showed themselves to be remarkably adaptable to the changed conditions of their history.

In spite of this vigorous capacity in both races for assimilation, the Indians' successes drew the envy and enmity of the more vociferous and politically ambitious white settlers. When the country began to fill up, as the backwoodsman would say, the attack on the Indians' presence took on the guise of a defense of the settlers' interests. In her book *Davy Crockett* (1934), Constance Rourke pointed out that some whites thought the Indians must be very rich: invariably, they seemed to live on the best soil for growing cotton, they were said to occupy hills full of gold, and they controlled regions which were surely more blessed with water and game. Her opinion of the president's role in this affair was that "Perhaps as commanding general in the Creek War Jackson had lost sympathy with the Indians, even though friendly Creeks and Cherokees had served him well. Political pressure doubtless played a part" (p. 134). It seems that Andrew Jackson, no better or worse a man than most politicians, a man whose chief support came from powerful leaders in the West and South who promoted America's manifest destiny, now considered the Indians an impediment to the expanding nation's primary economic interests.

Ironically, it had been Jackson himself who had resolved the Creek War by concluding a treaty promising the Creeks ownership of and protection upon their lands—forever. Today, the long, sad story of governmental betrayal of the Indian treaties throughout the eighteenth and nineteenth centuries is no longer a suppressed paradox of American history, and Jackson's role in this matter cannot be called an anomaly. Crockett's resistance, however, is an anomaly: it is not to be expected, it diverges from the norm, and it has nothing to do with progress and politics. Incredibly, we find our frontier hero siding in this issue with New England reformers and speaking of the noble savage as if he had read and believed James Fenimore Cooper. Perhaps this is because he had help in writing his speech from easterners who made a profession out of building the images of politicians. But much in his speech and in his response to the situation bears the authentic stamp of Crockett's independent character.

What Crockett said in Congress on that day in May seems at first glance to be inconsistent with some contemptuous remarks about Indians which he made nearly four years later in his autobiography. But, as we will see, the Indians in his book are the enemy he had fought years earlier in the Creek War, not his neighbors in West Tennessee. Moreover, the contents of the speech as we have it are consistent with his lifelong opinion of the thoughtless mechanisms of an expanding government and are perfectly representative of his characteristic habit of favoring the underdog in any fight. He spoke as a man of conscience, which he was:

> He had his constituents to settle with, he was aware; and should like to please them as well as other gentlemen; but he had also a settlement to make at the bar of his God; and what his conscience dictated to be just and right he would do, be the consequences what they might.
> (Blair, *Davy Crockett*, p. 211)

He said his constituents expected him to live by his own vision of his duty and to compel the government to do its duty by keeping its commitments. He was, sadly, wrong in this point: his constituency favored the bill. He argued against what everybody else declared to be a necessary "solution" to an inescapable "problem," because he knew that the perceptions and definitions promoted by any political group are inherently suspect. While he realized that the Indian removal bill was unjust, his larger concern was that its specific features would not really work to favor white settlers in the areas from which the Indians would be dislocated. The bill did not meet his test of common sense, and it was his lively display of common sense that had given him a good reputation as a local magistrate and Tennessee state representative.

Since he was not in actuality a reformer, let alone a political philosopher, Crockett did not defend the Indians' rights as a matter of abstract morality. He said in the speech that if the Indians would agree to move and if the plan provided them a fair exchange of lands, he would vote for such a bill. He seems to have been most interested in the Indians' courageous behavior and their stubborn refusal to cooperate. He understood their reluctance to believe the government's promises; he did not trust the government, either. What Crockett admired was the Indians' firm-willed resistance. They had expressed their defiance in brave words, he said: " 'No: we will take death here at our homes. Let them come and tomahawk us here at home: we are willing to die, but never to remove.' " His respect for their attempt to stand against the blind, impersonal force of a distant authority is characteristic of Crockett. He could see that the Indians were in a position analogous to that of the poor white settlers who had opened West Tennessee to agriculture and trade and who were now losing their homes. Landowners in North Carolina, who had laid claim earlier to the indefinite territories extending west of their holdings, were moving in to take the farms so painfully scratched out of the forest by the unlettered and politically naive "squatters." Crockett had already lost a battle over the settlers' rights to the land they worked, and he had paid as dearly for his opposition to Jackson on that issue as he would now for opposing Indian removal.

In both fights, Crockett displayed that peculiar backwoods self-reliance which defined itself not so much by principle as by opposition to any and all assumptions the system took for granted. He characteristically resisted pressure from any quarter, but became especially uncooperative when Congress considered measures which expressed the will of powerful politicians and rich speculators who lived in older communities or on established plantations far away from the western lands and who knew nothing of personal labor upon the soil. In the closing chapter of his autobiography, he recalls his opposition to the Indian removal bill in terms which suggest that he believed he could be true to himself only if he remained conscientiously independent of Jackson:

> His famous, or rather I should say his in-*famous*, Indian bill was brought forward, and I opposed it from the purest motives in the world. Several of my colleagues got around me, and told me how well they loved me, and that I was ruining myself. They said this was a favourite measure of the president, and I ought to go for it. I told them I believed it was a wicked, unjust measure, and that I should go against it, let the cost to myself be what it might; that I was willing to go with General Jackson in every thing that I believed was honest and right; but, further than this, I wouldn't go for him, or any other man

> in the whole creation; that I would sooner be honestly and politically d---nd, than hypocritically immortalized.
>
> (*N*, pp. 205-6)

Crockett spoke against the bill out of suspicion and from bitter experience. The five Indian nations would lose to Jackson's government just as the settlers were losing to the eastern landholders, and a defeat for the Indians might make it easier for the absentee claimants to force the pioneer families out of their cabins and off the farms they had made. Crockett must have known that the settlers themselves did not see it this way. But he admired the brave resistance of both the Indian and the pioneer, and he felt a close bond with any person threatened with the loss of what he had built with his own hands. It had happened to him and would again. What is consistent in Crockett's public life is his defense of the concrete and immediate concerns of the people who lived on the land against the abstract and distant claims imposed by people who lived according to the artificial constructions of warrant and law.

The historian who wrote the definitive modern biography of Crockett, the late James A. Shackford, argues that Crockett's political naiveté, so attractive to the backwoods and small-town voters and so much a part of his legend, continually worked against him. Apparently he did not realize that his Indian removal speech would be reported in the *Jackson Gazette* and included in the collection which appeared later. The anthology was published by eastern political factions who used Crockett's speech as part of their appeal to those who might sympathize with the Indians in their plight. Professor Shackford points out that publicizing his speech was "part of the Whig plan" to build Crockett into "an anti-Jacksonite of national proportions" (*DC*, p. 116). Other politicians were voraciously exploiting the public's image of Davy Crockett, and he became entangled in the impossible task of trying to live up to his own legend. He tried to keep the news of his position out of Tennessee and away from those voters who wanted the Indians removed; at the same time, he was trying to be the man who liked to say, "Be always sure you're right—then go ahead!"

It was characteristic of Crockett to take a stand against some powerful authority and then tell everyone who pretended to represent the official position to go straight to hell. His objection to letting the Jacksonites employ unconstrained means in the execution of a dubious project is a measure of his backwoods common sense. The abstractions he opposed were power and authority, forces which would use the people's money as a weapon against their better interests. At the end of his autobiography, Crockett writes:

> I am at liberty to vote as my conscience and judgment dictates [sic] to be right, without the yoke of any party on me, or the driver at my

heels, with his whip in hand, commanding me to ge-wo-haw, just at his pleasure. Look at my arms, you will find no party hand-cuff on them! Look at my neck, you will not find there any collar, with the engraving

> MY DOG.
>   Andrew Jackson.

He meant what he said, and whether or not he always lived up to this declaration of his independence, it represented the style which made common folk love him and upon which his legend as an American type is built. His cantankerous, satirical attitude caused his nearly absolute failure to accomplish anything at all in the United States Congress. His militant individualism was an ideal which did not serve his political purposes or his personal hopes and which, in general, has never been the sort of stance that changes the conditions of history. But it is this ideal that the inheritors of the legend have admired and preserved.

Moreover, the stories constituting Crockett's legend reflect the independent style of his life and of his talk. He was inventive, self-reliant, irreverent, and above all humorous. Indeed, the dominant characteristic of his legend is its amplification of his humor rather than his achievements. Curiously, the story of how the legend grew has its own humor, for virtually all of the artifacts of the Crockett legend are deliberate fabrications. The legendary Crockett is a collective invention which the real Crockett initiated and which his audience has perpetuated.

1

DAVID CROCKETT— THE FACTS

Paradoxically, the politically unfortunate combination of gullibility and pride in Crockett's character contributed to events which led to the fortunate publication of his one book-length contribution to American comic literature, his autobiography, *A Narrative of the Life of David Crockett of the State of Tennessee*, published in 1834 (hereafter referred to as his *Narrative*). With technical assistance from a talented friend, Crockett wrote this splendid account of his life partly in reaction to the appearance, in 1833, of a bogus biography most likely written by Mathew St. Clair Clarke, who was clerk of the House of Representatives. Clarke's book was first published in Cincinnati as *The Life and Adventures of Colonel David Crockett of West Tennessee* and later that year in New York and London as *Sketches and Eccentricities of Colonel David Crockett of West Tennessee*. (Clarke's book is hereafter referred to as *Sketches*.) Before the publication of James Shackford's *David Crockett* in 1956, Clarke's *Sketches* was widely accepted as a Crockett biography and wrongly attributed to James Strange French, a minor novelist from Virginia. Shackford makes a fair case for Clarke's authorship and a stronger one against the possibility that French wrote the book. French received the royalties, however, and his name apparently provided a cover for Clarke's political intentions—which are very obscure but may have been related to the Whigs' growing interest in Crockett's split with the Jacksonian Democrats. Although it is certainly an important part of the immense body of story and lore based on Crockett's life, and although Crockett probably told many of its anecdotes originally, the book itself is not biographical, nor was it published with Crockett's approval and assistance (*N*, p. xii; *DC*, pp. 258-64; Arpad, *A Narrative*, p. 23; Exman, p. 41).

Clarke's *Sketches* is written in an inflated, stylized prose, a sort of imitation literary language often found in the popular books of the early nineteenth century. The author talks extensively about Crockett's jokes, his "eccentricities," his backwoods manners, and his naiveté. He had known Crockett for some time, and he retells some of Crockett's yarns, trying awkwardly to capture his storytelling style. The result of using artificial literary refinements to frame the details of Crockett's life, character, and talk is that the book seems to mock the congressman even when the author tries to praise him. Crockett himself had probably told many of the book's stories to Clarke, but he was embarrassed when he saw the final product in print. He declared publicly that the book was untrue and that he did not know who its author was, and he repeated his disclaimers in the preface to his own autobiography, the *Narrative*. But Clarke's *Sketches* had already become the source of a good many legendary and sometimes ludicrous anecdotes that were widely retold and reprinted. Crockett thus continually met stories about himself coming back to him in forms which surprised and sometimes appalled him. He naturally felt obliged to set the record straight and to present his own stories in a plain, unvarnished style suitable to a man from West Tennessee.

Crockett's desire to dispel the distortions perpetrated by Clarke's *Sketches* was one motivation for writing the *Narrative*, but the actual creative work, which took place during a frenzied few weeks in January and February of 1834, was precipitated when an interesting political opportunity presented itself. Crockett's name had been mentioned wherever there was speculation about the candidates for the 1836 presidential election. Jackson had been reelected in 1832, and the Democrats were already promoting Martin Van Buren as a successor. In December 1833, the Whigs, whose ranks contained quite a few disenchanted former Democrats, approached Crockett, who had long been fiercely at odds with Jackson, to suggest that he cross over and let them consider running him against Van Buren (see *DC*, pp. 143-44). Now the publication of a good self-portrait became an imperative. The appropriate format lay at hand: the campaign biography or autobiography was a widely used political device and a well-established prose genre. Crockett protests so much in his *Narrative* that he is not interested in the presidency, he makes it seem certain that he is. Curiously, his own book contains some information that is in Clarke's *Sketches*, and this served further to promote the proliferation of legendary yarns about him. He was discovering that his legend had a powerful life of its own, and that even he could no longer retrieve all the underlying facts.

THE ART OF AUTOBIOGRAPHY

In producing the *Narrative*, Crockett enjoyed the collaboration of a skilled writer, Thomas Chilton, a congressman from Kentucky. The text itself stands as the evidence of Chilton and Crockett's distinctive achievement, and it shows that the Kentuckian shared the Tennessean's way of talking and understood his storytelling style. Stanley Folmsbee, who edited for publication James Shackford's scholarly edition (1973) of Crockett's *Narrative*, believed that the book was essentially Crockett's in terms of its content and Chilton's in terms of its language (*N*, p. xvi). However, as the selections from the *Narrative* in chapter 3 of this volume show, the book's language is artistically true to the style of both Crockett and his public image, the legendary Davy. Chilton and Crockett exercised just enough artifice to make their portrayal of Crockett's hyperbolic talk and lively storytelling convincing. Being from contiguous states within a single geographical region, they knew the same backwoods usages and figures of speech, and they worked them into the book judiciously, selecting those most likely to be widely perceived as authentic. A more accurate assessment than Folmsbee's would assign the technical details of writing to Chilton, the stories to Crockett, and the book's credible language to the partnership.

The special backwoods idioms are embedded in a first-person narration addressed in a familiar tone to an implied reader who is assumed to be an honest, practical, and open-minded fellow. As the narrator, Crockett is posed as a friendly, modest, straight-shooting common man. He speaks in perfect confidence, as if he knew the reader to be a sympathetic listener. He says that he became a politician only because he was compelled by popular acclaim and a sense of duty to take a position of leadership. The narrator's language is a fully developed model, or artificial representation, of the backwoods or rural vernacular. It is consistent in its deviations from standard English, and its idioms are authentic, which is readily demonstrated by pointing out that most of them have survived in vernacular American speech.

Crockett's pronunciations and rhythms are suggested rather than literally duplicated in every detail. The art which makes the language credible is not forced by excessive misspelling, strained metaphor, or esoteric diction. These last three devices are the sure indicators of an inept or inexperienced writer who implicitly asks his reader to remain aloof from his dialect-speaking persona. The storytelling style and content of the *Narrative* are thus distinctly Crockett's. The subtler techniques by which it achieves an illusion of authenticity, the most important of which would have been restraint, are probably the effect of Chilton's good judgment. The reader who is not a dialect specialist is more likely to identify the book's style as being generally representative of American backwoods tall talk and less likely to associate it with a definite region or even a specific class of folks.

It is the book's finely managed art which invites comparison with Benjamin Franklin's *Autobiography*, which was first published in America in 1818. History records that Crockett owned an 1825 edition and had signed his name in it (Arpad, *A Narrative*, p. 29). Like Franklin's, Crockett's autobiography stresses the facts most useful to an idea of the man's whole attitude toward life, rather than pretending to display all the facts of his life. Each author omits mentioning his serious faults, like Franklin's callous treatment of his wife or Crockett's bad habit of missing congressional roll calls, but each happily recalls a few harmless mistakes in order to illustrate how much he has learned from sad experience.

Both Franklin and Crockett declare their moral purposes, insisting that they have an obligation to instruct their readers in the fundamental public values—honest political behavior, common sense, hard work, and pursuit of the common good. Franklin says he was tempted to title his autobiography *The Art of Virtue*, and his list of the thirteen essential virtues is the highlight of the book; Crockett leads off with his famous epigraph, "Be always sure you're right—then go ahead!" and declares that his motive is a desire for justice, not fame. Both life stories are presented as paradigms illustrating the American ideals of individualism and self-reliance. Both narrators cultivate aphorisms and anecdotes, and neither takes himself too seriously.

Indeed, the form of both autobiographies is so clearly invented and their details so carefully reconstructed that either might be thought of as a humorous didactic novel. The episodes are derived from each author's history, but they are enlarged and made significant by his talent for truthful fiction. In both, the narrator is a cheerful, brilliant eccentric who unselfconsciously promotes inventiveness, independence, and optimism. This is why the two autobiographies tell us so much about each man's comic imagination, even though they give us only some of the historical details and omit anything that would badly damage the image being created before the reader's very eyes.

There is, however, a large and telling difference between the two autobiographies. Franklin knew his life served as a great example, so he selected a few incidents which would illustrate his many accomplishments and wove these into a rhetorical design showing that he had used his freedom very well. He encourages young Americans to do likewise. Franklin was quite literally a self-made man, and his autobiography is a deliberate re-creation of that original self-creation. He began to write in 1771, nineteen years before his death; he left off the account of his life with 1759, and never finished the book. So it is that Franklin's autobiography scarcely begins to enumerate his many and varied successes. Instead, it portrays just one of his many protean selves, the witty, persevering, rational, and honest man who seems wholly incapable of subterfuge. Crockett's autobiography, on the other hand, contains nearly all of the historically important events of his life up through his last term in Congress, and ends with January 1834, only two years and two months before his death at the Alamo. Much art went into the language of Crockett's recollections and the rhetoric embellishing his image, but it must not have been difficult to decide which successes to include and which failures to exclude. Crockett's *Narrative* is distinguished by its style, not its historical content.

Crockett may have contributed little to the history of American political institutions, but even before he wrote his book, his yarn-spinner's talent and his ability to articulate the hopes and values of the independent backwoodsman had already initiated the rapid processes which transformed him into a legendary figure. And he appeared at just the right historical moment: the type of the heroic American backwoodsman was already established in the public mind. Crockett fitted the role perfectly, and he played it so well that he made it all his own. Early in his public career, Crockett understood that he was becoming a legend, and he found this peculiar fame puzzling as well as satisfying. In the preface to his *Narrative*, he states that he is writing to correct the misconceptions perpetrated by the spurious biography of the preceding year, and then adds:

> I have also been operated on by another consideration. It is this:—I know, that obscure as I am, my name is making considerable deal of fuss in the world. I can't tell why it is, nor in what it is to end. Go where I will, everybody seems anxious to get a peep at me. . . . There must therefore be something in me, or about me, that attracts attention, which is even mysterious to myself. I can't understand it, and I therefore put all the facts down, leaving the reader free to take his choice of them.
>
> (N, p. 7)

This should not be dismissed simply as an expression of "extreme and naive egotism," as Stanley Folmsbee has said (*N*, p. 7, fn. 8). Folmsbee is recalling the famous portrayal of a "vain, ignorant Davy Crockett" in *Main Currents in American Thought*, where V. L. Parrington mistakenly labels the Crockett legend a "myth" and uses the word to mean "lie" (vol. 2, pp. 172-79). The note of genuine perplexity in Crockett's paragraph suggests instead that his self-assessment is simply a true perception of his curious and sometimes disturbing fame.

Crockett was painfully aware of how easily his folksy, unlettered manner could be parodied. He was always surprised and overly pleased by admiration and support, and other politicians could take advantage of his naiveté by flattering him or arranging some public honor. He was so seldom rewarded with political success that any measure of adulation tended to distort his judgment. It seems natural for him to have played up his image: he was a born showman, a yarn spinner, an actor. At the end of the preface, he tells us that another of his motives for writing an autobiography is to clarify his public role: "But just read for yourself, and my ears for a heel tap, if before you get through you don't say, with many a good-natured smile and hearty laugh, 'This is truly the very thing itself—the exact image of its Author' " (*N*, pp. 10-11). A heel tap is the drop of whiskey left in a shot glass after it has been emptied, and a straight shot of whiskey is the traditional device for saluting the successful completion of a yarn spun among friends.

Crockett's contemporary reader would have entered the *Narrative* in much the same way he prepared to see a strolling actor, watch a stage comedian, hear a political orator, or participate in a joke and story session: on guard, willing to play the game, and with disbelief suspended for the moment. In spite of his painful awareness of his naiveté, Crockett often managed to transform his supposed limitations into advantages by using his backwoods style in a productive mockery of itself. In a campaign debate, he could win over a crowd by pretending not to understand an opponent's complicated argument or schoolbook vocabulary. Here, in his preface, he says that he has no "pretension" to grammar and dismisses spelling by declaring that it is "not my trade." These gestures are designed to engage the sympathies of a large audience of workers, farmers, and settlers. The backwoods style of self-deprecation deflates the pretensions of educated opponents and defuses the volatile self-consciousness of unlettered folks. Implicitly, Crockett invites them to take his side in the game of common sense against book learning. He can safely expect his reader to know that whenever he modestly says he is surprised that an ordinary man like himself has been mentioned as a candidate, he is really saying he considers himself a prime candidate. Crockett's genius is displayed throughout his *Narrative* as a clear understanding of how best to evoke his audience's sense of values. Besides all this, the book is humorous and entertaining in its own right—a valuable work of comic art and an exemplary artifact of popular culture.

Since most of the important events of Crockett's life are included in the *Narrative*, it also serves as the basic document of the formal record, providing the outline for any scholarly biography. For his biography, James Shackford searched out numerous other sources: letters, official papers, memoirs by persons associated with Crockett, periodicals of the time, Tennessee and Texas state archives, the historical files of several county offices, congressional records, and genealogies. Professor Shackford's biography is scrupulously faithful to this record and emphasizes actual events rather than the voluminous popular lore which has distracted other biographers, virtually all of whom found themselves more engaged by the legend than by the facts. The book expands and illuminates Crockett's autobiography, but it cannot make of him an important figure of American political history. Paradoxically, no historian would pursue the facts of Crockett's life if it were not for his legend, and many have concluded their research by contributing to the legendary lore rather than clarifying those facts. Professor Shackford understood this, and he did a splendid job of setting Crockett's life in the historical context so that we could see the matrix from which the legend rose.

A BIRTHRIGHT OF POVERTY

Both the historian and the student of popular culture must find the beginning of Crockett's story satisfying, for it exemplifies perfectly both the history and the legend of the American westward migration. His family's history first emerges to our notice from the murky background of violent confrontation between the earliest pioneer settlers and the Indians who lived west of the Appalachian mountain barrier. The name of Crockett's grandfather, David, is written in the Lincolnton, North Carolina, court house records for 1771. Four years later, he and his family crossed the Appalachians into the area which is now the northeastern tip of Tennessee. Potential settlers and speculators considered this region part of the western lands reserved for the state of North Carolina; at the same time, it was one of the last areas assigned by treaty to the Cherokee and Creek nations, after a succession of broken treaties and boundary setbacks. In 1776, pioneers living at the Watauga Settlement, an independent community inside the disputed territory, fought several skirmishes with the Indians. Finding their location untenable, they sought to obtain official organization and protection through annexation to North Carolina. Their petition to the state legislature was signed by David Crockett, the grandfather, and his son William. In 1777, yet another treaty established a new Indian boundary line and enabled settlers to purchase land, but by 1778, North Carolina was compelled to pass a law forbidding whites to survey or even travel through the territory (*DC*, pp. 3-5).

By then the evolution of hostilities was well advanced, and the Indians, in desperation, were trying to uproot the numerous settlements already planted west of their boundary. The Crockett homestead was one of these. In the spring of 1777, an Indian raiding party killed the first David Crockett and his wife. In the *Narrative*, the grandson writes, "By the Creeks, my grandfather and grandmother Crockett were both murdered, in their own house, and on the very spot of ground where Rogersville, in Hawkins county, now stands" (*N*, p. 15). At home with their parents were Joseph and James, Crockett's uncles. A shot broke Joseph's arm, and James, who could neither hear nor speak, was taken prisoner. Crockett says that James lived with the Indians for nearly eighteen years and was "delivered up" from his bondage by being "purchased" from "an Indian trader" (*N*, p. 16). James Shackford adds that "For years thereafter he tried unsuccessfully to relocate gold and silver mines he had visited blindfolded while a captive of the tribes" (*DC*, p. 5). Other brothers in the family were not at home when the raid occurred; one, John, was serving as a ranger defending the frontier outposts, and it is he who became the father of the legendary Crockett.

Bernard DeVoto painted a vivid picture of frontier life in *Mark Twain's America* (1932), a book which opens by referring to the appearance in 1835 of Halley's comet and the commission from Old Hickory to Colonel

Crockett to scale the Alleghenies and wring the tail off it. It was the year Sam Clemens was born. Important information illuminating these opening scenes in the story of Crockett's life is provided by DeVoto's sharp detailing of the painful side of the pioneer adventure. DeVoto's book does not neglect the glories and joys of life in the woods, but over the years it has provided a corrective to the constant exaggerations which characterize our sentimental fictions. DeVoto reminds Americans that the settler in the backwoods was an unpredictable man who had most likely been a failure at more civilized pursuits. The settler and his wife would not have sought the dangers and discomforts of homesteading in a wilderness if they had enjoyed some more promising opportunity back home. DeVoto writes, "These people, being rejected, were too poor to buy or make shoes, and their privies were the naked earth. Medical science had not yet studied the etiology of the hookworm disease" (p. 55). The evil air—*mala aria*—and a host of other, as yet unnamed, diseases collectively called "fever" shared with the weather an awesome absolute power over the lives of individuals. Check any man, woman, or child living on the frontier at any time and "A clinical thermometer would have shown about one degree of fever" (DeVoto, p. 57). Likewise, James Shackford writes, "Poverty, as well as danger, was the birthright of the pioneer; and John Crockett inherited his full share of it" (*DC*, p. 5).

In the *Narrative*, Crockett treats his family's poverty humorously, never mentioning it as an affliction, but simply stating it as a fact. He is, of course, portraying himself as a self-made man; he knows that the story of his poverty enhances his image because so many of his constituents could tell an identical tale: "As my father was very poor, and living as he did *far back in the back woods*, he had neither the means nor the opportunity to give me, or any of the rest of his children, any learning" (*N*, p. 16). Crockett's earliest recollection includes the detail that he was not yet old enough to have any "knowledge of the use of breeches, for I had never had any nor worn any" (*N*, p. 18).

John Crockett's life was marked by experimentation and plenty of bad luck. In the familiar pattern of pioneer family life, David's father can be seen searching, moving from one settlement to another, never quite finding prosperity or stability. He married Rebecca Hawkins in 1780; they had three daughters and six sons, of which David Crockett was the fifth. The size of the family is typical, and in a strictly economic accounting of the settlers' lives, this number of children can be seen as both asset and liability: if there were many mouths to feed, there were also more hands to work. Women paid for the losses shown on this kind of balance sheet. But John Crockett was a responsible man. He had fought in the Revolution at the battle of Kings Mountain in North Carolina, he held office several times in various places as a constable or magistrate, and he always worked hard to support his family. Many backwoodsmen living in America at the end of the eighteenth century were illiterate, but Rebecca and John Crockett were not. Both signed documents with their names, not their marks, and John easily managed the legal details required of his offices. Offsetting these indications ever so slightly are clues to the man's limitations and misfortunes: several failures at farming, the loss of a homestead through bankruptcy, and a curious charge of petty larceny, of which he was found not guilty. Shackford notes also that John and Rebecca speculated in land, but gained nothing by it (*DC*, p. 6 and fn. 10).

QUICK WITS AND FAST FEET

David Crockett was born August 17, 1786, "at the mouth of Lime Stone, on the Nola-chucky river," in what is now Greene County, east of Knoxville. The probable site is marked today by a marble slab (*N*, p. 17, fn. 13; Holbrook, p. 170). In 1794 John Crockett went partners with Thomas Galbreath to build a mill on Cove Creek in Greene County. Emblematic of John's usual luck, this venture was literally washed out, as David says, "when there came the second epistle to Noah's fresh, and away went their mill, shot, lock, and barrel" (*N*, p. 21). John then moved his family to Jefferson County, where he put up a rough log building on the road between Knoxville and Abingdon, Virginia, and opened a tavern. Crockett says the tavern allowed only the barest of livings:

> His tavern was on a small scale, as he was poor; and the principal accommodations which he kept, were for the waggoners who travelled the road. Here I remained with him until I was twelve years old; and about that time, you may guess, if you belong to Yankee land, or reckon, if like me you belong to the back-woods, that I began to make up my acquaintance with hard times, and a plenty of them.
>
> (*N*, p. 22)

The year was 1796, the same in which Tennessee, after an abortive attempt to become the state of Franklin, was admitted to the Union—the sixteenth state, and the first to be created out of government territory.

If frontier children were to be assets, they had to be workers. Crockett's father hired him out to a man named Jacob Siler, who was moving to Rockbridge, Virginia, and had stopped at the tavern. He was driving a herd of cattle, and he pressed John Crockett to provide him with an assistant. David was scarcely twelve years old, and it was winter. In the *Narrative*, he forgives his father in a matter-of-fact tone, recognizing poverty's imperatives and adding a fine humorous twist which evokes his political purposes:

> Being hard run every way, and having no thought, as I believe, that I was cut out for a Congressman or the like, young as I was, and as little as I knew about travelling, or being from home, he hired me to the old Dutchman, to go four hundred miles on foot, with a perfect stranger.
>
> (*N*, pp. 22-23)

We may be reminded that when Benjamin Franklin was twelve, his father bound him in service to an older brother, a disagreeable arrangement which led to his breaking all family ties to begin his travels in that paradoxical condition of poverty and freedom which he took to be the perfect opportunity for the exercise of self-reliance.

But on this trip, young Crockett was to learn mainly that some people take advantage of children when they can. In return for his pains, he may have discovered hidden veins of fortitude and resourcefulness in himself. Siler pretended to be kind to him and paid him several dollars at the end of their journey, but then he tried first to persuade and next to force David to stay with him instead of returning home. David pretended to be satisfied, and worked for several weeks, partly because he thought his father might have expected him to obey Siler and partly because he was biding his time. When three wagoners came by headed toward Knoxville, David recognized them as occasional visitors to his father's tavern. He quietly arranged to meet them far down the road the next morning. It was bitterly cold when he awoke three hours before dawn. The boy trudged seven miles through deep snow to join the wagoners and start toward Tennessee. Fortunately, his escape and the trip home taught young David that people can also be kind, for strangers helped him on the way.

Crockett had no formal schooling until he was eighteen, when he hired himself out to a teacher named Kennedy and for six months worked two days a week in exchange for four days of reading, writing, and "cyphering." In his *Narrative*, he tells a funny story about how an earlier attempt was made to educate him. David was thirteen in the fall of 1799 when his father "took it into his head" to send him to a little country school which was taught by Benjamin Kitchen. Before the end of the first week, during which he had just started "to learn my letters a little," he had an argument with another boy in the schoolroom. After school he lay in ambush and pitched into his enemy when he came by. Afraid the master would lick him in turn, David pretended for several days to go to school, but played hookey instead. One morning, David's father received a note from Schoolmaster Kitchen inquiring about the boy's absences. In a rage, John Crockett set out to drive young David to school with a stick.

Another interesting glimpse into backwoods family life is provided by Crockett's remark that his father had taken a few "horns" of whiskey and "was in a good condition to make the fur fly." Whiskey was a staple, and many a man took some of "the creature" to start his day, but more crucial here is the signal that Crockett's father perpetrated the common confusion between discipline and physical punishment. Vowing to whip him harder than the master would have, David's father cut a two-year old hickory sapling—something more than a switch—and started after him:

> We had a tolerable tough race for about a mile; but mind me, not on the school-house road, for I was trying to get as far the t'other way as

possible. And I yet believe, if my father and the schoolmaster could both have levied on me about that time, I should never have been called on to sit in the councils of the nation, for I think they would have used me up.

(*N*, p. 31)

The size of the hickory his father had cut impressed David considerably.

Escape led directly to young Crockett's running away, the commencement of a journey that would last for more than two years and take him into manhood. Perhaps this interlude provided him with a more useful education than the schoolmaster intended. Perhaps it is of some significance to his legend that this was a distinctly picaresque journey: surviving by virtue of his quick wits and fast feet, taking advantage of opportunities as they arose, he lived from one episode to the next. He herded cattle into Virginia, hired out at day labor on farms, and drove wagons to Baltimore. There, in the spring of 1800, David stood on the wharf and dreamed of shipping out to London. He even made an agreement to sign on with one of the captains, but the wagoner for whom he was working at the time would not release him. Again David made his escape, but not to the sea; instead, he had the luck to meet with a more generous wagon master who was headed toward Tennessee.

But home was far away in terms of time and distractions. David worked for one farmer or driver after another for nearly two more years. It was February or March of 1802 before he made his famous crossing of the New River in the southwestern tip of Virginia. The river was very wide there, and an icy storm was whipping up whitecaps. David was forced two miles downstream before he could land on the other bank. His canoe was swamped and he was freezing, but finally he made his way to a farmhouse where, as Crockett writes, " 'a leetle of the creater,'—that warmer of the cold, and cooler of the hot," made him right again (*N*, p. 42). In Sullivan County, Tennessee, he came to the house of his uncle, Joseph Crockett, where he stayed a few weeks.

When David arrived home at last, he was moved to try a sentimental deception. It was nearly dark, and he said little, talking only enough to secure his lodging for the night. "I had been gone so long, and had grown so much, that the family did not at first know me. And another, and perhaps a stronger reason was, they had no thought or expectation of me, for they all [had] long given me up for finally lost." David lay back until all the tavern guests were called to supper. "We had sat down to the table and begun to eat, when my eldest sister recollected me; she sprung up, ran and seized me around the neck, and exclaimed, 'Here is my lost brother.' " Then Crockett adds, in words that sound exactly like one of Huck Finn's lessons in the relationship between love and guilt, "The joy of my sisters and my mother, and, indeed, of all the family, was such that it humbled me, and made me sorry that I hadn't submitted to a hundred whippings, sooner than cause so much affliction as they had suffered on my account" (*N*, pp. 42-43).

GETTING "MADE UP" IN LIFE

John Crockett was never out of debt, and David tells stories about how, during the next two or three years after his return, he worked off some of the notes his father had accrued—six months' work for Abraham Wilson to pay off thirty-six dollars, six months for the Quaker farmer John Kennedy to pay off forty dollars. After bringing Kennedy's note to his father, Crockett went back to work in that household and stayed until summer 1803. His first infatuation seems to have occurred at about this time. The farmer's niece came for a visit, and young David was smitten. She was already engaged, however, and to one of John Kennedy's sons. Disappointed, Crockett surveyed the situation and decided that "all my misfortunes growed out of my want of learning" (*N*, p. 49). He gave up any thought of pursuing this particular young woman, but his observation led him to seek the schooling referred to earlier, his six months with Kennedy's married son, who lived not far away.

Before long, Crockett met a family named Elder which boasted several pretty daughters, and he fell in love with Margaret. As coincidence would have it, he and Margaret served as attendants at the wedding of Kennedy's son and niece, and the event fired Crockett's intention to marry. He worked hard to win Margaret, but she was evasive: "I would have agreed to fight a whole regiment of wild cats if she would only have said she would have me" (*N*, p. 50). In a moment of weakness, Margaret did say yes, but before long, their marriage plans collapsed.

In his *Narrative*, Crockett says Margaret Elder deceived him. Other evidence suggests that she was understandably offended by his conduct. On a Saturday several days before the appointed wedding day, he set out to visit her, intending to stop at a shooting match along the way. The proud possessor of a very good new rifle, Crockett won the whole beef and sold it for five dollars "in the real grit" (hard money, not banknotes), and the victory, he recalls, put him in a "flow of good humour." Shackford finds, however, that Crockett is not telling the whole truth about that good humor. The shooting match and the carousing growing out of it evolved rapidly into a regular backwoods frolic, and it lasted into Sunday. Much the worse for wear, Crockett finally arrived at the house of Margaret's uncle, about two miles from her family's place. There he met her sister, who sobbed out the news that Margaret intended to marry another young man. Doubtless, Margaret had by now heard the stories about David's marksmanship and his capacity for celebration, and probably she considered his tardiness just one more example of his casual regard for respectable behavior. All along, Crockett had been reluctant to inform her family about the specific date of their marriage or to discuss with them the details of the ceremony. Now she had good cause to remedy the mistake she had made in her moment of weakness.

Interestingly enough, some of the details which reflect badly on Crockett's behavior in this incident are in the spurious biography of 1833, but not in Crockett's *Narrative*. Shackford confirms the version to be found in Clarke's *Sketches* and points out that Crockett's own life history is revisionist with regard to the story of Margaret Elder. Of course, the contrast between the two accounts throws the whole episode into sharp relief and allows us a clear view of one of Crockett and Chilton's artifices. The portrait of the irresponsible young carouser in Clarke's *Sketches* is transformed into an image of the good-hearted, vulnerable man of sentiment in the *Narrative*. Although he reports that the whole affair literally made him ill with "the worst kind of sickness,—a sickness of the heart, and all the tender parts," Crockett was soon cured (*N*, p. 55). He was young—nineteen—and had neither time for, nor disposition toward, melancholy. Within less than a year, he had courted and won Polly Finley.

Crockett's story about his courtship of Polly and their wedding is rich with interesting details about country customs and attitudes. He was convinced that a young man was not "made up" in life until he had a wife and family. David declared his eligibility and intentions all around the neighborhood, and several other persons took an interest in his quest for a wife. The complicated, highly ordered business of matchmaking had begun. A smart young woman whom David liked but would not court—"she was as ugly as a stone fence"—told him she knew of a very eligible girl, and arranged a meeting for him with Polly's mother. Mrs. Finley, a spirited, talkative woman, immediately began to play the game, teasing David and calling him her "son-in-law."

It was arranged that David and Polly would meet at a reaping, a three-day community frolic which included dancing for the young people, skits and games for the children, and partying for everyone. Polly had made herself scarce before David arrived at the frolic, but Mrs. Finley greeted him heartily and opened the courtship herself:

> Her mamma, however, was no way bashful. She came up to me, and began to praise my red cheeks, and said she had a sweetheart for me. I had no doubt she had been told what I come for, and all about it. In the evening I was introduced to her daughter, and I must confess, I was plaguy well pleased with her from the word go. She had a good countenance, and was very pretty, and I was full bent on making up an acquaintance with her.
>
> <div align="right">(<i>N</i>, p. 59)</div>

David and Polly danced and talked nearly all the night, and in the morning took parts for the children's entertainments. Crockett realized very soon that his best chance lay with her mother: "I went on the old saying, of salting the cow to catch the calf" (*N*, p. 60). When the party was over, David returned to his place of employment and began to bargain with the younger Kennedy about working in barter for a low-priced horse. He had begun to assume his familial responsibilities.

Things went well enough for a time, and David was especially pleased with Polly's father, "a very clever old man." In this context, "clever" means "congenial" (*N*, p. 61, fn. 5). After a while, however, a rival suitor turned up, and Mrs. Finley, to everyone's surprise, no longer favored David. David sought an explanation from Polly and found out that she, at least, still liked him best. He confronted his rival, and the two young men competed fiercely—and comically—for Polly's full attention. One would corner her and talk while the other fumed; then the other would cut in and let the first stew for a while. This went on hour after hour until finally the intruder went off "gritting his teeth with pure disappointment."

Mrs. Finley remained obstinate, but then an incident occurred which clinched the courtship of David and Polly. Crockett had joined a number of other men in a big wolf hunt, and while riding through a strange woods, he became separated from the party. It grew dark overhead, and he writes that he became "scared" of how "it clouded up"—that is, anxious to move on because he could see that he was about to be caught in the storm. Then, after riding six or seven miles, he realized he was hopelessly lost. Suddenly he saw a woman running through the trees and quickly caught up with her. It turned out to be none other than Polly, who had been in the woods hunting for her father's strayed horses. This chance meeting was delightful to both, but finding one another also meant that they were now lost together, and they were dead tired. Fortunately, just before dark, they came across a path which led to a house. Often a life-or-death matter, the first law of the frontier was hospitality, and the family gave them shelter and food. Crockett records that, tired though they were, they "set up all night courting; and in the morning we parted" (*N*, pp. 60-64).

Now Crockett was more than ever eager to marry, and the next time they met, he and Polly set their wedding date. His first duty was to arrange with his father for an infare—a reception and celebration in which the groom's family welcomes the bride into their fold. But when David went to her home to talk about the Finleys' part of the ceremonies, Polly's mother exploded in a rage. Crockett writes, "she looked at me as savage as a meat axe. The old man appeared quite willing, and treated me very clever. But I hadn't been there long, before the old woman as good as ordered me out of her house." Crockett gently reminded her of her earlier good disposition toward him, but "her Irish was up too high to do any thing with her, and so I quit trying. All I cared for was, to have her daughter on my side." Feeling insulted, David arranged with Polly that he would come for her the following Thursday, and if her mother still refused to cooperate, they would ride away to be married by a justice of the peace (*N*, pp. 64-65).

According to custom, a groom would gather his party together and ride to the bride's house. They would wait at the gate while the best man and two or three attendants entered the house to present an empty flask. As a token of his acceptance, the bride's father would fill the flask with his best, and the return of this symbol to the groom signaled the commencement of ceremonies. This excellent ritual took on a rich new significance in the case of David and Polly, for on this particular Thursday, there seemed every chance that the flask would come back empty. The party met a wrathful mother when they entered the house, but the congenial Mr. Finley ignored her and filled the flask. The attendants hastily retreated to return it to David. Instead of drinking from the flask, Crockett rode up to the door, leading a second horse saddled for Polly. He was ready to elope if Mrs. Finley should resist again, but Mr. Finley took his wife aside and persuaded her to accept the inevitable. He even got her to apologize to David. Crockett emptied the flask, and the celebration was conducted as was right—in the bride's home and in a receptive atmosphere. But before long, cold reality reestablished its claims. Crockett humorously records, "I thought I was completely made up, and needed nothing more in the whole world. But I soon found this was all a mistake—for now having a wife, I wanted every thing else; and, worse than all, I had nothing to give for it" (*N*, p. 67). Perhaps the truth of his absolute poverty is what had set Polly's mother against him, but he does not admit that he came to perceive her side of the controversy.

It is precisely at this point that we see Crockett beginning an endless search, caught up in that westward migration which seemed to offer new hope to so many desperately poor Americans. The assumption of responsibility compelled him to seek a better living. Sharecropping on rented land would not do: "In this time we had two sons, and I found I was better at increasing my family than my fortune" (*N*, p. 68). But in addition, as Shackford says, "he was riding out upon an impulse of the race" (*DC*, p. 17). Crockett chose to move to Lincoln County on the southern edge of Middle Tennessee, into country drained by the Duck and Elk Rivers. Although he certainly understood such things by the time he wrote his book, he could not have known when he set out in 1811 that the elements of legend were embedded in this frontier opportunity. It was no longer a chance to be taken by just a few hardy and reckless souls; it had become a popular myth, and thousands upon thousands of people were moving west. Carving out a living in the woods measured men and women exactly: their survival, let alone their prosperity, was tied directly to their ingenuity and energy. Until old laws were enforced to assign ownership—a calamity which was soon enough to catch up with the Tennessee pioneers—the land was perceived to be open, and squatter's rights prevailed for the time.

Wild animals were a critical factor. Many a family, Crockett's included, survived their first seasons without crops because there was plenty of game in the woods. To this day, throughout the South from Kentucky to the Gulf

Coast, autumn hunts are called harvests. American backwoods men and women might be thought of as the last hunters and gatherers. Ironically, the Indians they displaced had lived as such for millenia, while the civilization which followed the pioneers has reduced to nearly zero the land's capacity for sustaining human life naturally, without agriculture. Many farm families depended upon game even after their crops were established, since the land and the markets were not always generous to corn and cotton. Crockett was pleased with the new country insofar as he could kill an "abundance" of deer and smaller game. But significantly, "the bear had been much hunted in those parts before, and were not so plenty as I could have wished" (*N*, p. 69).

It is difficult for modern Americans to conceive of a way of life in which the extermination of large predators was deemed a private necessity or a public service. The wolf hunt mentioned earlier was not a sporting event, but a cooperative effort to purge the countryside of animals which threatened livestock. The farmers saw themselves as working in direct competition with wolves, bears, and mountain lions, which in the lowlands were called panthers or painters. Today we mourn the disappearance of these animals and make feeble attempts to define narrow territories where a few might roam without interference. In Crockett's time, the presence of wolves was too immediate for comfort. More than one mother living in a remote log cabin won local fame by shooting a panther that was stalking her chickens or her children.

Bears, however, were much more than a threat. Bears were a staple of life. A fine, fat bear provided several barrels of meat—something like pork—which could be salted down for the winter, an excellent hide for a warm blanket or coat, and many pounds of lard for cooking, making soap, waterproofing leather, and lubricating tools and wheels. A growing family like Crockett's could have used half a dozen a year. There was no way a frontiersman ever could have anticipated the day when such an awesome and useful creature would have vanished from the woods. Actually, the bears vanished because the woods vanished; hunting itself, Crockett's remark notwithstanding, did not account for all the loss. Most important to this legend, at the center of the various portraits of America's frontier heroes is the dynamic picture of the bear and the hunter locked in mortal combat, tooth against knife. This grim and hilarious drama was neither tragedy nor comedy, and the literature which sprang from it took form as the violent tall tale told by the humorous storyteller. Of all his roles, Crockett was always most comfortable in these two—bear hunter and yarn spinner.

INDIANS!

In 1813 Crockett and his family backed up a little and moved east into Franklin County, where they established themselves on a branch of Bean's

Creek, about ten miles southwest of Winchester and five or six miles north of the Alabama line. White settlers in this area and in the lands to the west and south were actually in violation of a treaty with the Creek Indians, who were now, under the leadership of Tecumseh, trying to resist the tide. Several skirmishes and what the whites called "massacres" had occurred in Franklin County just a few months before Crockett arrived, and this may help explain why he was so anxious to join the fray. The fact that the British were allied with the Indians may also have contributed to his decision. Curiously, however, Crockett's enlistment in the militia was most directly motivated by news of a bloody fiasco which had taken place far away, at Fort Mims, three hundred miles south of his new home. Certainly his neighborhood was not immediately threatened.

Fort Mims was located near the junction of the Tombigbee and Alabama rivers, just north of Mobile. Its mission was to defend settlements along the major waterways of South Alabama. Ironically, many of the settlers there were children of intermarriages between whites and Indians. In August 1813, some 260 soldiers and about three hundred other persons were sheltered in the fort, which was commanded by Major Daniel Beasley of the Mississippi Volunteers. The United States government's interests in the area are clear enough. Acquisition of the disputed lands, by a war of attrition against the Creeks, was the obvious though unofficial purpose; another objective, to interfere with the alliance between the English and the Indians, was part of the overall strategy for the 1812 war. In addition, the Spanish, who had not yielded to the argument that West Florida was part of the Louisiana Purchase, were supplying the Indians with arms through the important ports of Mobile and Pensacola. As Shackford points out, the Creek Indian War can be thought of as having its beginning, not with the Fort Mims massacre in August of 1813, but earlier in the summer when soldiers attacked a band of Indians moving north from Pensacola carrying guns, ammunition, and other supplies. A party of about 180 recruits surprised the Indians near the banks of Burnt Corn Creek, where they had stopped to rest in the shade. The Indians were routed, but the soldiers, instead of exploiting their advantage, stopped to loot the supplies that had been left behind. Though outnumbered two to one, the Indians regrouped and counterattacked, and they did not repeat the soldiers' tactical error by failing to follow through (*DC*, p. 18).

This ludicrous humiliation was disaster enough, but worse was yet to come. At Fort Mims, Major Beasley had assured General Claiborne that he was ready for anything, but in fact his preparations had been minimal. On August 30, a small army of Creeks found the gate at Fort Mims open and the guard positions unmanned. Officers were playing cards and some of the soldiers were dancing with young girls from the families who had sought refuge there. Ironically again, the Creeks were under the command of a peacemaker—William Weatherford, an Indian who was friendly to the settlers and whose secret purpose was to avoid a bloody confrontation. He had come to the fort fully expecting to find the gate locked. Apparently, he had intended to lead the attackers away after making a token, face-saving foray. When the Indians found Major Beasley frantically trying to close the gate against the mound of sand it was scooping up ahead of itself, Weatherford's hopes suddenly became irrelevant. Beasley fell first. Weatherford pulled back, but found himself all alone; the thousand warriors with him stormed the fort and overwhelmed the garrison. Five hundred men, women, and children were killed; only fifty or sixty survived—many of these, slaves who were taken as part of the spoils of war.

In the *Narrative*, Crockett has a long and intriguing paragraph explaining why he was so eager to go fight the Indians. It is filled with the standard rhetoric which distinguishes all justifications of the Indian wars: "The Creek Indians had commenced their open hostilities by a most bloody butchery . . . my countrymen had been murdered, and I knew that the next thing would be, that the Indians would be scalping the women and children all about there, if we didn't put a stop to it." It is impossible to tell whether "there" refers to southern Alabama or to Franklin County, Tennessee. Polly saw no crisis, and she pleaded with him not to go, but Crockett could not be dissuaded. The local militia was organizing, although they had no official order to do so: Jackson would not receive the news of Fort Mims until September 12, several days after his messenger had passed through Winchester and alerted the people in Crockett's neighborhood. War fever was upon them, and perhaps Crockett tells us what he really wants us to know at the very end of his paragraph when he candidly admits, "The truth is, my dander was up, and nothing but war could bring it right again" (*N*, pp. 71-73; see *DC*, pp. 18-21).

This hair-trigger eagerness to fight was typical of the frontiersman, but can scarcely be thought of as belonging exclusively either to the past or to America. It is distinctly masculine, and seems characteristic of men of action who prefer rapid change to the orderly processes of domesticity, law, and dialogue. It was an attitude that would come to be strongly associated with the legendary figure of Crockett, for it seems to have suggested to Americans the utility of prompt resolutions as opposed to the vague compromises of thoughtful diplomacy. Perhaps it was an attitude which served a purpose on the frontier, but proves merely destructive in a city.

To be sure, Crockett's ready willingness to make war upon the Indians in 1813 is one of those details which cast doubt on the sincerity of his plea against the Indian removal bill in 1830. But Crockett's earlier enthusiasm simply reflected the convictions of a whole society. Except for a minority of reformers, white Americans—especially westerners—supported the government's policies, which were ultimately to yield tragedies like those at Sand Creek in 1864 and Wounded Knee in 1890. In the matter of assessing Crockett's achievements, it might be best to recognize that any man can be of two minds, especially when circumstances change and seventeen years pass. Crockett should be credited with a capacity for moral growth: by 1830, he was well ahead of his contemporaries in his awareness of the injustices being perpetrated in the name of Manifest Destiny. As for his remark, "my dander was up," Crockett can be likened to Tom Sawyer in *Huckleberry Finn*. Perhaps if asked why he went, he might have given and likewise left unfinished the reply, "Why, I wanted the adventure of it; and I'd a waded neck-deep in blood to . . . ."

Crockett's company mustered in Winchester and then marched to Beaty's Spring, south of Huntsville. Here volunteers gathered until they numbered about thirteen hundred. Crockett comments, "I verily believe the whole army was of the real grit." But while waiting for this muster to be completed and for Jackson to come from Nashville with his regulars, Crockett had the opportunity to join a small party of the "best woodsmen, and such as were best with a rifle," in a scouting expedition under an officer named Gibson (*N*, pp. 74-75).

Gibson split his party south of the Tennessee River, taking one group with him and assigning the other to Crockett. There was a full moon, and Crockett's squad traveled mostly at night. Before long, they had gathered some hints of the Creeks' whereabouts, so they moved during the day to the crossroads where they expected to meet the other party. Gibson failed to rendevous. Crockett waited for him that evening, all night, and part of the next morning. His men were anxious to return to the main camp, but he convinced them that their duty was to go ahead until they had gotten what they came for—the whole picture of the Creeks' movements. Taking the initiative, Crockett followed up several leads until he found himself, after sundown, in the camp of some friendly Indians. There he kept his ears open while he competed with several young braves shooting arrows by the light of a pine knot. He soon learned that a large contingent of hostile Creeks was marching north to meet Jackson's army. Creeks on the warpath were called Red Sticks, so nicknamed because of the clubs they carried, which were painted to look as if they were smeared with the blood of the enemy. Crockett thought there was a good chance that a "whole nation" of Red Sticks might surprise the troops at muster; but much to his chagrin, when he reported back to Colonel John Coffee, his news was ignored. A day later, Gibson, who had been given up for lost, finally straggled in, bearing much the same intelligence. His report, of course, was promptly acted upon.

That the officer's word was believed when Crockett's was not taught Crockett "one of the hateful ways of the world." But he kept his tongue, even "though I was so mad that I was burning inside like a tarkiln, and I wonder that the smoke hadn't been pouring out of me at all points" (*N*, pp. 81-82). Crockett instantly perceived the ironic limitations of a military hierarchy: it is a strictly divided class system which automatically interferes with the mission itself. The army regulars, he believed, had their own special interests at heart, and the safety of the people would always come second. His experiences in Jackson's army and his astute observations as a common soldier would one day lead him to argue in Congress against appropriations for West Point and against supporting a regular army at the expense of maintaining a ready militia.

In response to Gibson's reports, Colonel Coffee sent the message on to General Jackson and ordered his troops to throw up a quarter-mile-long breastwork. "Old Hickory-face," as Crockett calls him here, was by now only a day away; he made a forced march and arrived the evening of October 10. The breastwork was never defended, for as soon as his own troops were rested, Jackson wisely set out to intercept the Indians further to the south. In battle, offense is nearly always superior to defense. Jackson split his army, and Crockett's contingent, abut eight hundred men under Coffee's command, crossed the Tennessee River, passed Huntsville, and waded the river again at Muscle Shoals. A number of horses were lost in the ford when their hooves became jammed in the rocks. Essentially, the troops were retracing the route of Crockett's reconnaissance. They soon came upon a place called the Black Warrior's Town, on the Black Warrior River, not far from the present location of Tuscaloosa (*N*, pp. 82-83). The Indians were not at home. The troops looted the town, gathering up all the dried beans and corn. They were glad to get the food: Crockett reports several times in his book that the army carried few supplies and did not manage to forage very well. His own tracking skill and quick rifle provided his messmates more than one meal. Having liberated what they wanted, the soldiers burned the Indian town to the ground. It was but an emblem of what was to come.

The two columns rejoined, and Jackson, having learned that the Red Sticks were camped at the Ten Islands, near what is now Gadsden, turned his army to the southeast toward the Coosa River. Some distance below an Indian town called Tallusahatchee, Jackson began to build a fortified camp, to be named Fort Strother, and from there he sent out scouts. The scouts returned to report that the town upriver was occupied by Creeks. It was Jackson's chance to avenge Fort Mims; and on November 3, 1813, he took full advantage of his opportunity.

The curious rhetoric of Crockett's report is mixed, revealing that he was both appalled and fascinated. There is some predictable racist diction in his telling. Describing the death of an Indian boy of about twelve, Crockett observes that the youth's arm and thigh were broken, and that he was crawling along so near a burning house, "the grease was stewing out of him." The young man made no cry, and Crockett says, "So sullen is the Indian, when his dander is up, that he had sooner die than make a noise, or ask for quarters" (*N*, p. 89). Still, Crockett's account also shows a streak of hard irony, a tone which tells us bluntly that he clearly perceived the mindless brutality of the whole affair. In the traditional manner of the frontier storyteller, he lets the sordid details speak for themselves, sentimentalizes nothing, and never gives cruelty the name of heroism. He does not seem to relish the tale of violence he is telling, and we understand that his purpose is to let us see the scene as it was.

The army formed a square around the town, and Hammond's rangers were sent in to precipitate the fight. The warriors could have fought from their homes, but probably they did not want to draw heavy fire toward their families, so they returned the rangers' volley and then pursued them, falling into the trap. Some of the Indians saw that they were surrounded, and tried to surrender. A few were taken prisoner, but most of them were shot down even as they dropped their weapons. The official reports do not portray the engagement as a massacre, and it is interesting that the detail of the Indians' attempt to surrender is usually omitted (*DC*, p. 25; *N*, pp. 88-89). As the army closed in, the Indians retreated to the houses. Women and children joined the battle. One woman sat in a doorway holding a bow with her feet and fired an arrow which killed a man named Moore. Enraged, the troops fired on her in turn, and she "had at least twenty balls blown through her." Then Crockett adds, "We now shot them like dogs; and then set the house on fire, and burned it up with the forty-six warriors in it" (*N*, p. 88).

This was the pattern of the whole attack. When it was over, nearly two hundred people had been shot or burned to death in the houses, and about eighty were prisoners. Crockett's ironic tone as he reports a subsequent event speaks for itself:

> No provisions had yet reached us, and we had now been for several days on half rations. However we went back to our Indian town on the next day, when many of the carcasses of the Indians were still to be seen. They looked very awful, for the burning had not entirely consumed them, but given them a very terrible appearance, at least what remained of them. It was, somehow or other, found out that the house had a potatoe cellar under it, and an immediate examination was made, for we were all as hungry as wolves. We found a fine chance of potatoes in it, and hunger compelled us to eat them, though I had a little rather not, if I could have helped it, for the oil of the Indians we had burned up on the day before had run down on them, and they looked like they had been stewed with fat meat.
>
> (*N*, pp. 89-90)

The horror of the scene is posed as being both factual and insane. The yarn spinner speaks literally, and any reader inclined to romantic fantasies about life in the western woods would be irredeemably disillusioned by this simple rhetoric.

Thirty miles further down the Coosa River, a large force of Creeks had laid siege to Fort Talladega, which was occupied by other Creek Indians who were friendly to the whites. The purpose of the siege appears to have been to force the Indians in the fort to join the fight against Jackson's army.

The chief of the friendly Creeks disguised himself in the skin of a hog, escaped from the fort, and made his way to Jackson's camp with the news (*N*, p. 91, fn. 19). Jackson marched his troops southwest early in the morning of November 7 and prepared to employ a tactic similar to the one that had worked so well at Tallusahatchee. The hostile Creeks knew he was coming and withdrew from their siege to hide in the woods between the fort and a large stream. Jackson split the army into a pincer which would trap the Indians between its arms and against the stream at their backs. Two scouting companies, commanded by Major Russell and Captain Evans, were sent forward to discover the Creeks' hiding places and bring on the battle. The scouts could not see the Indians in ambush beneath the banks of the stream itself. They rode past the fort on their way toward the stream, and friendly Indians lining the battlements tried to signal warnings to them. When it became apparent that the officers did not understand their signals, several Indians scrambled down and ran out to head off their horses. This brought the companies to a halt, and the Indians hiding under the banks of the stream opened fire. The soldiers were not well within range, and so were able to retreat easily into the fort or back toward the army's lines.

Now the Indians poured out from behind the bank, over a thousand strong, and "They were all painted as red as scarlet, and were just as naked as they were born" (*N*, p. 92). The army's two lines held fairly well at first, and three or four hundred warriors died in the crossfire as they dodged back and forth from one arm of the pincer to the other, trying to find a way out. At last, some of the Indians regrouped and began an effective return fire. Surprised and confused, two or three companies of soldiers—Crockett says they were "drafted militia"—broke ranks and retreated. The Indians quickly took advantage of the weak spot in the line, breaking through to get behind the troops still fighting in the pincer, which soon collapsed. This turn of events forced the enraged and frustrated Jackson to retreat to Fort Strother while taking harassing fire in the rear of the column. In spite of this, the score of the battle was one-sided: Jackson's army had lost only seventeen men (*N*, p. 93, fns. 23, 24).

Had he been able to follow through, Jackson might have terminated the Creek Indian War then and there. As it was, an alternative presented itself later in November, when he was offered an opportunity for peace negotiations through the Hillabee faction of the Creek nation. In one of those appalling and stupid errors which are characteristic of the chaos of all warfare, East Tennessee troops under General James White, who was acting under orders from General John Cocke, attacked the Hillabee Indians in several towns and massacred them. These two officers were unaware of Jackson's negotiations, and the Hillabees were slaughtered just when Jackson had begun to believe he had things under control (*N*, p. 93, fn. 24). The Creeks assumed that Jackson was behind this inadvertent double cross, and the war plunged into a drawn-out death struggle which would not end until the Creeks were completely subjugated.

Judging their work substantially complete, the volunteers from Crockett's part of Tennessee decided that it was time to go home. As Crockett says, their clothing was in rags, they were starving, and their horses were worn out. The *Narrative* at this point reflects Crockett's political purposes rather than the storyteller's imperative: revisionism displaces straightforward recounting. Their sixty days' enlistment was up, Crockett says, but the fact is that the men had enlisted for ninety days. Crockett tells a tale about how the troops mutinied and called Jackson's bluff; Shackford more accurately reports that it was Jackson who called the troops' bluff and quelled the mutiny before it gathered steam. According to Crockett, the mutiny succeeded and the men went home without a shot being fired, but it was actually an act of political intervention on the part of Tennessee Governor Willie Blount that compelled Jackson to let them go. To top it all off, Crockett himself was an observer of the attempted mutiny rather than a participant in it. The record shows that he stayed for his full ninety days' enlistment and did not go home until December 24, 1813.

Why did Crockett alter the facts and elect not to tell that he had honored his commitment? Shackford writes that Crockett's political ambitions are the determining factor here. Crockett revised the mutiny story in order to portray himself as a man who had stood up to Jackson from the beginning. The mutineers met Jackson's troops at a bridge, and there, Crockett says,

> we heard the guards cocking their guns, and we did the same; just as we have had it in Congress, while the "government" regulars and the people's volunteers have all been setting their political triggers. But, after all, we marched boldly on, and not a gun was fired, nor a life lost.
>
> (*N*, p. 95)

The quoted word "government" is supposed to stand for Jackson the usurper of powers, and now Crockett hopes no one will be afraid to stand with him and face the general at "the executive bridge" in the upcoming political battles of the 1830s. Crockett goes on to describe battles that Jackson's forces fought at Emuckfau Creek on January 22, 1814, and at Enotachopco Creek on January 24, but the official record does not confirm that he was present at these engagements (*DC*, pp. 27-28). He was at home—he says in the *Narrative* that he was on furlough—during the terrible and decisive Battle of Horseshoe Bend in March and while Jackson negotiated his treaty with the Creeks in August.

Crockett enlisted again in September and served as a sergeant until March 1815. He was to participate in a nasty search-and-destroy mission against some scattered bands of Creeks who had refused to live by the treaty and had fled southward toward the Gulf of Mexico. This adventure would take him into the trackless wasteland of Northwest Florida and school him further in the dimensions of human treachery.

THE LABYRINTH

In the early fall of 1814, while Crockett was busy at home with Polly and their two children, Jackson began to prepare an expedition to Pensacola. That steamy Gulf port was under the jurisdiction of Spain, and the authorities there had given sanctuary to the Creeks who had refuted the treaty. The Spanish had also harbored several British ships, and now some three hundred British soldiers were working closely with the Indians. The British planned to use this combined force in the battles they anticipated at Mobile and New Orleans. By late October, Jackson—now a major general—was ready to attack Pensacola. In the month preceding, Crockett had enlisted with Major Russell's "Separate Battalion of Tennessee Mounted Gunmen." Polly remained unpersuaded. "Here again the entreaties of my wife were thrown in the way of my going, but all in vain; for I always had a way of just going ahead, at whatever I had a mind to" (*N*, p. 101). Crockett and Russell's battalion marched south as soon as they were ready, but they were too late to join the main army, and so arrived in Pensacola on November 8, the day after the battle.

The Spanish had not defended Pensacola vigorously, but they had managed to blow up Fort Barrancas before Jackson could take it. The British had begun a discreet withdrawal. The few Creeks who remained had scattered into the swamps east of the Escambia River, which flows out of Alabama and across Northwest Florida into the upper estuaries of Pensacola Bay. On the evening of November 8, Crockett and some of his companions strolled down to the foot of Pensacola's main street, which ran directly to the bay, and there they saw the British fleet preparing to pull out. They bought a bottle in the heart of the most notorious district in town, and after taking a horn or two, they went back to camp. Russell's battalion was joined with another, and this regiment was marched back into Alabama and then, several days later, ordered to return to the Florida swamps for a mopping-up operation. The task is perfectly described in Crockett's idiom as a mission "to kill up the Indians on the Scamby river" (*N*, p. 103). While the main body of Jackson's army headed west to Mobile and on to the Battle of New Orleans, Crockett and his comrades were chasing small groups of Indians through the bayous and basins of the most forbidding forest in the Southeast.

This setting was to provide a real challenge to Crockett's skill as a scout and hunter. Here low-lying hummocks and plains of sandy soil supporting pines and scrub oaks are laced with an endless net of interconnecting creeks and basins—shallow lakes formed where the river system widens into large brackish pools which lie at sea level, but are still upriver from the bay itself. The regiment set up camp on the west bank of the Escambia. Then Major Russell, sixteen scouts including Crockett, and eighty or so Chickasaw and Choctaw Indians struck out across the river and to the east. They soon came to a basin, which Crockett accurately describes as "a place where the whole country was covered with water, and looked like a sea." The woods and creeks made going around it nearly impossible, so they "just put in like so many spaniels, and waded on, sometimes up to our armpits, until we reached the pine hills, which made our distance through the water about a mile and a half" (N, p. 107). The water was bitterly cold. After warming themselves by a fire, the company moved out, keeping half a dozen Indian "spies" ahead of them as an advance guard.

Before long, the forward spies came back to report that they had discovered a Creek Indian camp. The Chickasaws and Choctaws painted themselves in preparation for battle, and then told Major Russell that he should be painted, too, since he was an officer. Russell accepted. Before the troops could reach the enemy camp, however, two of the Choctaw scouts came upon two Creeks who were out hunting their horses. They struck up a friendly conversation, pretending to have escaped from Jackson's army in Pensacola. The Creeks told them where a large encampment of their comrades was, up on the Conecuh River, a tributary of the Escambia. Then, after politely taking their leave, the Choctaws turned around and shot the Creeks. When Crockett arrived with his party, he saw that the Choctaws had beheaded their victims and that the other Indians were taking coup by striking the heads with their war clubs. Crockett was invited to join the ritual, which he did. Afterwards, the Indians all gathered around him and called him Warrior! Warrior! (N, p. 110). Thus did soldiers like Crockett and Russell find themselves truly getting into their work.

Finally Russell's scouts found their way to the Indian camp which had been reported first. They attacked and killed a number of warriors there. Crockett adds, "they took two squaws, and ten children, but killed none of them, of course." Later, another party under Major Russell, but not including Crockett, fought a skirmish and took several prisoners. These were sent under Indian guard to Fort Montgomery, but Crockett reports that he later heard that "the Indians killed and scalped all the prisoners.... I cannot positively say it was true, but I think it entirely probable, for it is very much like the Indian character" (N, pp. 111-13).

Next the regiment set out across the panhandle of Florida, crossing the Choctawhatchee River and then striking further east toward the Apalachicola. This country was, as it still is, rich in game. There were deer, wild hogs, squirrels, snapping turtles, turkeys, and fish and oysters in the brackish estuaries, but Crockett's company nearly starved. An abundance of wild animals meant nothing to men who did not know their way through the labyrinthine terrain. The soldiers hunted every day and were never choosy about what they killed. Crockett says he took "every hawk, bird, and squirrel that I could find" (N, p. 116; see chapter 3 of this volume for Crockett's wonderful account of this absurd adventure). At the end of each day, the hunters threw all their game into a pitifully small pile, and the meat was divided into shares. Of all the hunters, Crockett caught on to the special demands of this peculiar country the soonest, and after a while he began to score on turkeys and deer. Nonetheless, the troops fared hard until their mission in the swamps was over and they returned to Alabama and the relative comfort of Fort Strother. When his regiment was ordered out again, Crockett made arrangements for a substitute—a respectable and legal procedure—and gave "the balance of my wages" to a young man who was anxious to go fight Indians (*N*, p. 123). Thus he cut short his enlistment by about a month and went home to work his farm.

STARTING OVER AGAIN, AND AGAIN

The bliss of peace was short-lived. Early in 1815, the Crocketts' third child was born, their first girl. They named her Margaret, though she soon was nicknamed Polly like her mother, whose own given name was Mary. In the summer, for reasons which history has not recorded, Crockett

> met with the hardest trial which ever falls to the lot of man. Death, that cruel leveller of all distinctions,—to whom the prayers and tears of husbands, and of even helpless infancy, are addressed in vain,—entered my humble cottage, and tore from my children an affectionate good mother, and from me a tender and loving wife.
>
> (*N*, p. 125)

The exact date of Mary Crockett's death is not known, nor is the cause. Perhaps she died of complications arising from her third pregnancy, which would account for the popular notion that says she suffered a long time and declined slowly; but there is no hint of this in Crockett's account. It is equally possible that she was struck down by any one of the many rampant infections which plagued frontier families. Crockett buried Polly in the woods and marked her grave with a cairn of stones.

It is hard to admit that David Crockett ever met a challenge which was too much for him, but the truth is that a backwoods farm was often more of a burden than even a man, a woman, and several vigorous children could carry. He writes candidly that he did not fare well trying to be both father and mother. Crockett's youngest brother and his family moved in to help, but this arrangement did not work out. "So I came to the conclusion," writes Crockett, "it wouldn't do, but that I must have another wife" (*N*, p. 126).

Elizabeth Patton was a strong and intelligent woman whose husband, James, had been killed in the Creek War. She came from a prominent North Carolina family and was said to have a little money of her own. She had two children, a son and a daughter who, Crockett says, were about the same ages as his younger two. She and David were married in Franklin County during the spring or summer of 1816. Their union was a promising one, and it certainly solved several of Crockett's most immediate problems. As he writes in the *Narrative*, he found it easy to be congenial with Elizabeth's children, she took readily to being the mother of his, and they "had a second crop together" (p. 127).

Elizabeth Crockett appears to have had a much better idea of how to manage finances than David was ever able to command. She was well organized and sensible. For several years and in many different ways, her talents superbly complemented Crockett's strengths and balanced his eccentricities. Her determination and confidence saved him from despair one grim day when they lost a promising business they had managed to build up during the first five years of their marriage. This venture included a grist mill, a powder mill, and a distillery, all set squarely on the banks of Shoal Creek in Lawrence County, Tennessee. In the fall of 1821, as if some force wished to recapitulate John Crockett's washed-out luck in 1794, a flood swept the mills away and smashed through the distillery, wrecking all the machinery.

Crockett says in the *Narrative* that he lost about $3,000 in this disaster; it is likely that part of this, the original capital investment itself, had been provided by Elizabeth. Both of them worked in the business, but apparently it was she who was known as the miller, for Crockett was more often out hunting and, by this time, electioneering. But the point of this story is to reveal something further about Elizabeth's strength of character. When Crockett returned from the state legislature to the scene of the calamity, he was dismayed. In writing about it, he manages a fine pun, taken from the language of the whiskey-making craft, to describe his situation humorously: "I had, of course, to stop my distillery, as my grinding was broken up; and, indeed, I may say, that the misfortune just made a complete mash of me." He means that, where spirits had once risen from mash heating in the still, no spirits nor spirit would ever rise from the kind of mash the flood had made of his still and of his heart. But Elizabeth showed her grit. His "honest wife," he says, turned to him and advised, "'Just pay up, as long as you have a bit's worth in the world; and then every body will be satisfied, and we will scuffle for more.'" Sparked by her courage, they worked off what they had borrowed "and took a bran-fire new start" (*N*, pp. 144-45).

Later in 1816, not long after his marriage to Elizabeth, Crockett set out with three of his neighbors to explore new country for the purpose of finding better soil. He clearly went for the adventure of it, too, but does not explicitly say so. They traveled south into Alabama, more or less following the old trail to Fort Mims, but soon began to experience real misfortune. One of the party, an expert hunter and woodsman named Frazier, was bitten by a snake—probably a cottonmouth, which is an aggressive reptile less inclined to retreat from a man than most poisonous American snakes. Frazier had to be left in the care of some friends of Crockett's who had settled in the Creek country after the war, and the explorers went on, though sorely handicapped by the loss of his talents. One night while they were camping, their horses wandered off—an ominous loss for travelers in the woods. Crockett went in search of them, hiking for a whole day—but he exaggerates tremendously when he says he covered fifty miles. He did not find the horses, and his extreme fatigue led to an attack of some vicious illness, probably the malaria which was to afflict him all his life (*DC*, p. 36). He collapsed in the woods, where he was found by two Indians. They signed to him that he would surely die, but he asked if there was a cabin anywhere near. They signed that there was, and one of the Indians half-carried him there. A kind woman took him in and nursed him until the next day, when two of his neighbors happened to come by. They moved him to the house of Jesse Jones, where he was attended for two weeks while he lay dangerously ill.

The standard treatment was bloodletting, so Crockett must have had to call upon all his strength to survive this ordeal. He says that his decline was reversed by the administration of a patent medicine which brought on the crisis. Crockett always manages a sharp-edged humor when writing about his own troubles, and here he says that Mrs. Jones, "who had a bottle of Batesman's draps, thought if they killed me, I would only die any how, and so she would try it with me." Crockett drank the whole bottle, which caused him to sweat furiously all night. When he awoke the next morning, he asked Mrs. Jones for a drink of water. "This almost alarmed her, for she was looking every minute for me to die." Perhaps Batesman's drops contained, besides the inevitable high percentage of alcohol, some measure of quinine, long used as a nonspecific drug for fever, and possibly an agent contributing, along with the bloodletting, to Crockett's ensuing anemia. When he got home, he was "so pale, and so much reduced, that my face looked like it had been half soled with brown paper" (*N*, pp. 130-32).

Crockett follows this tale with another good joke which illustrates both his political reason for writing the *Narrative* and his satirical skill:

> And I can't, for my life, help laughing now, to think, that when all my folks get around me, wanting good fat offices, how so many of them will say, "What a good thing it was that that kind woman had the bottle of draps, that saved PRESIDENT CROCKETT'S life,—the second greatest and best"!
>
> (*N*, p. 131)

The public knew that Jackson flagrantly cultivated a questionable spoils system, but always called him the "greatest and best" anyway. The jab is aimed at Jackson, but it also lets his reader think of Crockett as being good-natured about Jackson's supposed knavery.

Crockett returned home "to the utter astonishment" of his wife, for his companions had gotten back days earlier "and they reported that they had seen men who had helped to bury me; and who saw me draw my last breath. I know'd this was a whapper of a lie, as soon as I heard it" (*N*, p. 132). This nicely turned understatement springs from the same tradition as does Mark Twain's "the reports of my death have been grossly exaggerated" (see *DC*, p. 37). Crockett soon recovered, and again he resolved to find a new place to settle. In September of 1816, the territory east of the northward bend of the Tennessee River, in the south central part of the state, had been opened through a treaty with the Chickasaws; late in the fall, Crockett set out to explore it. He had ridden only about eighty miles from home when he was again interrupted by an attack of malaria. This time it was not bad enough to knock him down, so while he recuperated, he looked around a little and found himself "well pleased" with the Shoal Creek country. The Crocketts moved there in the spring of 1817 (*N*, p. 133; see fns. 19 and 20).

LAW AND ORDER—CROCKETT STYLE

What he says next not only illuminates Crockett's independent character, but identifies one of the central ideas of his legend: "It was just only a little distance in the purchase, and no order had been established there; but I thought I could get along without order as well as any body else." Idyllic pioneering "without order" always proved very temporary on the frontier, however, and Crockett's new neighborhood soon began to get a little crowded: "so many bad characters began to flock in upon us, that we found it necessary to set up a sort of temporary government of our own" (*N*, p. 133). The settlers met, and Crockett was chosen, unofficially, for his first position of public responsibility. As a backwoods magistrate making judgments without the benefit of lawbooks and legislature, he began to build his reputation for fairness and common sense. As he says, his word to the constable was as good as any warrant, and a judgment from him meant a debt soon collected.

Within a few months, the Tennessee legislature annexed the territory which was to become Lawrence County, and Crockett was officially made justice of the peace. Nonetheless, law on the books did not constitute law in fact, and order in the Shoal Creek country depended for two or three years upon Crockett's judgments, which always "stuck like wax, as I gave my decisions on the principles of common justice and honesty between man and man, and relied on natural born sense, and not on law learning to guide me; for I had never read a page in a law book in all my life" (*N*, p. 135).

The southern title of colonel, so facetiously used today, represented in Crockett's time an actual and substantial office. As in other states, Tennessee law required each county to maintain a militia regiment which elected its officers, kept itself in training, and stood ready to be called up for any emergency. The colonel would be the regiment's commander in time of war and would retain his title, if not his duties, when he moved elsewhere. A neighbor of Crockett's, Captain Matthews, who had been farming in the county a long time and who "made rather more corn than the rest of us," was running for colonel, and he asked Crockett to run for first major. Matthews gave a big frolic to kick off the campaign. During the festivities, Crockett learned that he had been set up: Matthews had asked him to run in order to gain his support, but had not told him that he would be running for first major against the younger Matthews, the captain's son. Clearly, Matthews did not believe that Crockett could win, but he made the mistake of saying that his son would rather run against anyone other than Crockett. Crockett cheerfully responded by telling Matthews that he need not worry, for he would run instead for the position of colonel. Later in the day, Matthews made his speech and informed the crowd that Crockett would now be his opponent. Crockett got on the stump next and told what had happened, "remarking that as I had the whole family to run against any way, I was determined to levy on the head of the mess."

The electorate was undoubtedly much amused by Crockett's straightforwardness and pure cheek, for he was elected colonel. He carried the title proudly all his life. Later, when he gained national fame, newspapers would generally refer to the man himself as Colonel Crockett or simply as Crockett; the name Davy Crockett seems to have become more closely associated with the legendary figure of the almanacs. Having told the comic story of how he became a colonel, Crockett marks the event in the *Narrative* as the beginning of his political career: "I just now began to take a rise, as in a little time I was asked to offer for the Legislature in the counties of Lawrence and Hickman" (*N*, p. 138).

Crockett writes that he knew nothing of government matters and had never seen a public document in his life. These claims are not literally true, but the sense of what he says reveals accurately another kind of truth: he conducted his first campaign intuitively, appealing to his constituents' faith in the ideal of folk wisdom, which is a strong and important American myth. He won because he let the people see that he was good-natured about his backwoods ignorance and actually thought of it as an asset. They understood that his open-mindedness more than compensated for his supposed limitations. In writing about his first campaign, he not only pretends he was forced to assume this humble pose, he recapitulates it, this time deliberately posing humbly in the pages of the *Narrative* while hinting that he could be called upon to run for the presidency. His opponent for the state legislature back then, he says, "didn't think, for a moment, that he was in any danger from an ignorant back-woods bear hunter. But I found I couldn't get off, and so I determined just to go ahead, and leave it to chance what I should say."

This approach worked well for Crockett. Whenever he rose to speak, he drew upon his ample store of anecdotes, relieving his embarrassment and winning the hearts of his audience by making fun of the whole political process and all its participants, including himself. He was, he says, like a fellow who was beating on an empty barrel because there had been some cider in it a few days earlier, "and he was trying to see if there was any then, but if there was he couldn't get at it. I told them that there had been a little bit of a speech in me a while ago, but I believed I couldn't get it out" (*N*, p. 141). This style always gained him great sympathy from ordinary folks who saw themselves as being wholesomely naive and who resented being looked down upon for their lack of fancy book-learning. To the present day, a great many Americans believe that unschooled common sense is a fundamental virtue which is easily impaired by too much reading or education. In David Crockett's character, unschooled common sense was in fact a noble and useful trait, but the schooling which would eventually impair this virtue in him is named politics.

Offsetting and supplementing the deliberate cultivation of his natural down-home humility, Crockett learned as he went along, and he learned fast. If he was rushed into some local phase of a campaign and did not know what the issues were, he would find out by listening carefully to the more experienced candidates' speeches. Sometimes, if he was forced to speak first at a gathering, he would tell several funny stories and then lead the crowd over to the refreshment stand for a general treat, leaving his opponent stranded on the stump. If he was to speak last, he took advantage of the fact that a crowd is always tired and bored after hearing too many long-winded speeches: he would rise, tell one good story, and quit. This ploy never failed to surprise and delight everyone. In the fall of 1820, he was elected to the state legislature by a wide margin, and he began his service in 1821. The following year was the year of the Crocketts' financial disaster, the loss of their mills and distillery, and their subsequent move to the Obion River country. Crockett was elected to the legislature a second time as a representative from this newly settled region, and he served until 1825.

BEARS!

Crockett's *Narrative* for this period devotes many more pages to his hunting adventures in the woods and canebrakes of West Tennessee than to his politics; thus it tells us much about the sources of his legend and little about his accomplishments as a public servant. This is our good fortune, for his stories are wonderful, while his accomplishments were mundane. The one great political battle of his whole career was his effort to secure a law which would allow the settlers clear title to the lands they had developed. The victory in this fight would never come to Crockett himself, but to his son, years after David's death at the Alamo. It is his style, not his deeds, which has engaged his audience, and from the beginning it seems as if Crockett's truest contributions to history are symbolic: he was becoming the image of a grand American idea, and historians who look for greatness in his concrete political acts are always disappointed.

The Obion River country lay within newly created Carroll County, an immense region that had been designated open to settlement in November of 1821 (*DC*, p. 53). The area was bounded by that part of the Tennessee River which flows north and by the Mississippi, and would one day be further divided into all the counties of West Tennessee. In 1822 Crockett and his family chose a homestead between the Rutherford Fork and the South Fork of the Obion, probably not far from the present site of the town of Rutherford in what is now Gibson County. When Crockett explored it during the winter following his first legislative session, the Obion River country was a paradise for deer, small game, and bears. The famous earthquakes of 1811 and 1812, which formed Reelfoot Lake and once caused the Mississippi to flow backwards, had here disrupted the rolling hills and valleys, scattering fissures across the face of the earth and felling enormous trees. (Three of the earthquakes in this series are estimated to have been the most massive in American history, each stronger by far than the 1906 San Francisco quake. See Johnston, p. 60.)

Not long after the earthquakes, a succession of fierce storms ripped across the same landscape, leveling whole forests. Consequently, Crockett calls the devastated region a "harricane." The word as he uses it may have two meanings. In the Midlands, any violent storm system was called a hurricane and would invariably have been a series of thunderstorms accompanied by tornadoes, rather than the giant circular storm which strikes the Atlantic and Gulf Coasts. A section of the woods which had been devastated and then covered by secondary growth was called a hurricane thicket.

The second meaning that Crockett may have intended has to do with a tall southern grass commonly called giant cane (*Arundinaria gigantea*). Fields of cane sprang up in the devastated areas, and for a while it would have competed fiercely with the new shrubs and trees that were striving to replace the original forest. A canebrake (in Middle English, "brake" means thicket) was thus a wonderfully tangled place. A hunter entering it would find himself among tall canes twelve to fourteen feet high. As he worked through the brake, he would find his way continuously blocked by logs, brambles, whole tops of dead trees, and thickets of saplings. A harricane, as Crockett spells it, would thus be a thicket of harried cane (Middle English "harien" means to lay waste or assault), and it could also mean a canebrake which harries the hunter who finds himself in it. The Middle English words in these combinations have survived in the vernacular still used and cherished by people living in the mountainous back country of eastern Tennessee, western North Carolina, and northern Georgia.

In the Obion country, the canebrake maze was further complicated and enriched by the crevices formed by the earthquakes. The lush new growth meant abundant food for all kinds of animals; the fallen trees and fresh gulleys meant cover. Crockett instantly recognized the significance of the fact that the Indians had continued to hunt here, and he understood perfectly the importance of such wealth to new settlers who would have to live off the land until their homestead was established.

Crockett explored the Obion country in the winter of early 1822, and on one occasion killed half a dozen whitetail bucks and two elk, all within a few hours. Later in the spring, he took ten black bears. With this wealth he repaid the hospitality he had enjoyed in the homes of the few families already settled there. That spring, with some help from his new neighbors, he put up a cabin and planted corn. The nearest homestead was seven miles away; the next nearest, fifteen. In the fall, Crockett brought his family to their new home, took in his crop, and harvested the deer and the bears. Crockett's accounts of his many hunts in the cane provide a striking record of the plentifulness and usefulness of these animals.

Given our historical and literary perspective, we can see in Crockett's bear-hunting yarns an ancient story pattern, one in which the hero confronts an archetypal, dangerous, manlike adversary, as threatening as Grendel in *Beowulf*, but quite real and perfectly edible. In America, the mythic and the practical often merged in precisely this way. On the one hand, Crockett poses humbly as a simple backwoods hunter; on the other, his compressed, humorous, and realistically detailed narrating of incredible adventures constitutes a highly developed art form, the formulaic bragging which is requisite to the hero's role. Keeping in mind the likelihood that his contemporaries would read his book partly for the purpose of judging his pragmatic character, Crockett always proceeds rapidly from tall-tale telling to straightforward remarks upon the economic utility of living off the land: "We got our meat home, and I had the pleasure to know that we now had plenty, and that of the best; and I continued through the winter to supply my family abundantly with bear-meat and venison from the woods" (*N*, pp. 164-65; see chapter 3 of this volume for some good samples of Crockett's bear-hunting stories).

IN THE PUBLIC INTEREST

Crockett's activities in the state legislature show that the pattern of his political career was established early. He would for the rest of his life find himself fighting and losing the battle for legislation that would provide a way for West Tennessee settlers to buy at a minimal price the farmlands they had so desperately carved out of the woods. He understood his cause well. Back in Lawrence County, his debtors were collecting every last cent from the sale of his and Elizabeth's holdings. The family literally came into the new country with no resources but their hands and their tools, of which the most valuable must have been Crockett's rifle.

In 1789 North Carolina had offered its western territory to the United States government for the purpose of creating a new state, with the provision that all the land warrants issued to North Carolina soldiers as a reward for service in the Revolutionary War would be honored. In 1806 Congress specified that the larger eastern portion of the state, more than two-thirds of Tennessee, would be set aside for the satisfaction of these original warrants. The western portion was supposed to remain open public land, but the eastern portion proved insufficient for satisfying all the warrants that suddenly appeared out of nowhere. Speculators had bought many of the original warrants from the North Carolina veterans or their heirs (Rourke, *Davy*, p. 130). In addition, postdating and the issuing of warrants in the names of dead persons were common practices. The woefully inaccurate records, when there were records, simply complicated matters. The fact is that no one knew how much territory would be needed to satisfy all the warrants. The most expedient solution was to use the public lands west of the congressional line (*DC*, pp. 48-49).

Over the years, in a series of attempts to resolve the matter, Tennessee kept trying to set cutoff dates after which no North Carolina warrants would be honored. Both the boundary line and the cutoff dates were continually set back, however, and by the time Crockett entered the fight, no settler in West Tennessee could hope to keep the land he had cleared; he might be dispossessed at any time by strangers bearing legal papers issued far away and long ago. To add to the irony, a West Tennessee settler usually found himself working hilly, rocky, and infertile soil, because the better lands, in Middle and East Tennessee, were long gone. Crockett's defense of the West Tennesseans' cause begins with his support of a move to call a Constitutional Convention. He hoped that a constitution would guarantee equal representation by specifying that the number of representatives from West Tennessee would be allowed to increase in proportion to its rapidly growing population.

Crockett also sought tax relief. As things stood, the poor lands of West Tennessee were taxed at the same rate as the rich plantations in Middle Tennessee (*DC*, pp. 48-51). The abstractions which mark the conflict are familiar ones: self-reliant pioneers fighting for their right to the fruits of their toil were pitted against strongly entrenched landowners fighting to preserve their hard-won vested interests.

A curious trait in the character of the pioneer is illustrated by Crockett's vote against a bill which would have outlawed gambling (*DC*, p. 52). The backwoods settler banked on his wits and his luck. He believed that liberty included the right to take risks against long odds in the quest for sudden wealth. Beneath all the legal language and political maneuvering, we can see that Crockett's position in history represents an era, a time in which a last-ditch effort was being made to maintain the deeper spirit that motivated the pioneer's awesome gamble. This spirit was the hope for independence which lay with a chance to acquire land without capital. It was this opportunity that had brought many Europeans to America in the first place, and now Crockett could see that before long this great idea would be an anachronism. The losers would be successive generations of native-born Americans. As it turned out, the frontier would remain open until the 1890s, but the shapes of those far-western opportunities were very vague in Crockett's time, or else had not yet appeared. Crockett's own search for a truly open frontier is the motif and motivation of his unsettled life, and it is the impulse which sent him eventually into Texas and to his fate.

A series of comic misadventures led to Crockett's second term in state politics. While the family worked hard to open their new farm, raccoon pelts, bearskins, and wolf scalps—which brought a whopping bounty of three dollars—furnished a cash crop. In February 1823, Crockett sold a load of skins in the town of Jackson. There, while taking a drink with some of his "old fellow-soldiers," he was kidded about the possibility of running for the legislature again. One of his companions was Dr. William Butler, Andrew Jackson's nephew-in-law. Crockett told them he could not realistically consider campaigning because he lived forty miles from any settlement and was too busy working the homestead. A week later, a hunter passing by the cabin surprised David and Elizabeth by showing them a newspaper clipping which announced Crockett's candidacy. David saw that the whole thing was a burlesque upon him, but it got his dander up, so he determined to go ahead with it simply to even the score (*N*, pp. 166-67).

Now Crockett put his experience to work, developing and exploiting the backwoods image which was so naturally his. He had someone make him a buckskin shirt with two oversized pockets big enough to hold a large plug of chewing tobacco in one and a pint of whiskey in the other. He knew that when he offered a voter a dram, the man would have to throw out his quid. After they had shared a little of the creature and talked a while, Crockett would offer the voter a chew to replace the one he had discarded. "And in this way he would not be worse off than when I found him; and I would be sure to leave him in a first-rate good humour" (*N*, p. 169). Rare is the politician who can claim to have done as well by the voters.

The parties who had twitted him in the first place soon realized that Crockett would be a real threat. The man they nominated to oppose him was Dr. Butler himself. Crockett used Butler's wealth against him, telling the voters that the doctor walked upon rugs in his home which were made of finer material than the voters' wives had ever worn. In those days, candidates for an office traveled together from debate to debate. Crockett had ample opportunity to memorize everything Butler said, so one day he asked to be first on the program. He rose and gave Butler's speech word for word, leaving his opponent literally speechless. Being a sharp politician himself, however, Butler did manage a spontaneous substitute; but Crockett's tactic carried the day (*DC*, p. 64). Crockett won the election with ease, and went to the legislature as the representative of five new counties in West Tennessee. There still had been no Constitutional Convention, so his district remained unequally represented. Perhaps Crockett's brash aggressiveness and good humor made up for the numerical deficit.

The biggest fight of the 1823 legislature led to Crockett's first serious public break with Jackson's supporters. The two state bodies met in a joint session to elect a United States senator. (Senators were not elected by popular vote until after the seventeenth amendment to the Constitution went into effect in 1913.) The incumbent was Colonel John Williams, an enemy of Jackson's since the Creek War. The Jackson forces were hard pressed to find a man strong enough to defeat Williams, and finally they persuaded Jackson himself to run, in spite of the fact that he was then campaigning for the presidency. Crockett greatly admired Williams, did not trust Jackson after the mutiny episode, and did not believe that Jackson's camp favored West Tennessee interests. He spoke out and voted for Williams, but Jackson was elected by a vote of thirty-five to twenty-five. Having secured the Senate seat for his party, Jackson later withdrew from the office (*DC*, p. 67). The day would come when Crockett, in the national Congress, would pay for his opposition to Jackson in this and other matters; but at this point in the *Narrative*, he simply cites the event as an early sign of his political independence: "I am more certain now that I was right than ever. I told the people it was the best vote I ever gave; that I had supported the public interest, and cleared my conscience in giving it, instead of gratifying the private ambition of a man" (*N*, p. 171).

The second of Crockett's terms in the state legislature continued through the usual two sessions and ended in October 1824. On at least three occasions, he valiantly supported the rights of the West Tennessee pioneers. He fought down a proposal to sell off a large portion of public lands for cash only, a measure that would have disenfranchised virtually every farmer who lived from crop to crop without ever accumulating a single dollar. He opposed a new measure to extend the time limits of North Carolina land warrants in West Tennessee. He lost this one. In November 1823, Crockett's vote helped pass a measure instructing Tennessee senators and congressmen to support a law in Washington which would allow Tennessee itself to dispose of vacant lands. Such a law might return some control over their fate to the farmers on the land. The sum of all his efforts would not be success: Crockett would go to the end of his political career without winning the pioneers' right to own their lands. However, the state legislature did call for a Constitutional Convention, the motion carrying by exactly the required two-thirds majority. Crockett's may well have been the deciding vote. By now he represented not five but ten counties in West Tennessee (*DC*, pp. 68-72).

As the second session drew to a close, Crockett was urged to run for the national Congress. The incumbent was Colonel Adam Alexander, a man of considerable means and experience. Alexander had worked to obtain a tariff law, and by coincidence cotton had sold in 1824 for the astronomical price of twenty-five dollars a hundred. Naturally, Alexander claimed credit for this temporary rise in the farmers' fortunes, and in the election of 1825 he narrowly eked out a victory over Crockett (*DC*, pp. 73-74). By 1826, however, cotton would be back down to six or eight dollars, and in 1827 the tide of political fortune would turn in Crockett's direction.

NECK OR NOTHING

Meanwhile, David and Elizabeth continued their struggle to gain some small measure of solvency. The record is dismal, and it remains so until Crockett's death. The years following the move into the Obion River country were marked by a number of futile attempts to expand the family's holdings—the purchase of a little land here, a business venture there. This meant that Crockett took on one debt after another, and occasionally he was sued in court for a settlement. It is possible that his expedition into Texas in 1835 and 1836 was the culmination of his efforts to get out of debt. He needed a big break and was looking for a homestead in a new country, when he rode instead into the absurd and dazzling page of history known as the Alamo.

One of those futile business ventures evolved into the biggest event and, as Crockett tells it, the most humorous episode of the time between his state legislative experience and his election to the national Congress. In the fall of 1825, Crockett hired some hands, and together he and his crew cut staves—the planks used to make barrels or wooden aquaducts—and built two rafts to carry them. They were going to sell both staves and rafts in New Orleans, but their departure was delayed throughout the winter while Crockett laid in bear meat for his family. He likewise accepted every invitation to help his neighbors lay in theirs. It is clear that Crockett preferred hunting to working, or considered hunting to be a man's true work: he devotes twenty pages of his *Narrative* to the hunts of this winter, and concludes by claiming to have killed 105 bears during the season. So it was that the crew and their two overloaded rafts did not leave Lake Obion until early spring of 1826.

Thirty thousand staves made the rafts top-heavy, and Crockett knew nothing of river navigation. The rafts got out of the Obion River without mishap, but when they entered the Mississippi, Crockett found out that his crew, including his pilot, knew no more about the big river than he did. The whole parade immediately began to drift sideways. Crockett thought it would help to lash the rafts together, and this they did; but now the boats were suddenly "so heavy and obstinate, that it was next akin to impossible to do any thing at all with them, or to guide them right in the river" (*N*, p. 195). Worse, Crockett found that they could not even force the rafts to run aground; they were bound to whatever fate the river held in store for them. Night fell, and every hazard was doubly dangerous because unseen. Crockett desperately searched for a way to rig some kind of effective steering, thinking all the while how superior bear hunting was to river rafting.

Disaster came in the dark, early morning hours. The strapped-together boats ran hard into a raft of logs which had piled up against an island, and immediately the first boat began to be sucked down, driven by the current like a wedge under the logs. Crockett was below deck. The only hole he could find was not quite large enough for him to squeeze through. As luck would have it, he stuck his head out above water, and there were his mates: "I told them I was sinking, and to pull my arms off, or force me through, for now I know'd well enough it was neck or nothing, come out or sink" (*N*, p. 198; see chapter 3 of this volume for Crockett's whole account of this wreck). He came out, of course, and by further good fortune, the whole crew survived, losing nothing more than their clothes and their cargo. A boat took them down to Memphis, and there Crockett met the postmaster, Marcus B. Winchester, who gave him and his friends some clothes and the price of their passage home. Out of this fiasco came a single good result: Winchester became one of Crockett's major supporters in the congressional election of 1827, backing him with both encouragement and cash (*DC*, pp. 78-79).

CROCKETT VS. WASHINGTON

Two shrewd, experienced politicians opposed Crockett in the 1827 election for Congress. They were Adam Alexander, the incumbent, and William Arnold, who was an attorney, a Jackson city commissioner, and a major general in the militia (*DC*, pp. 81-82; *N*, p. 201). All three happily indulged themselves in the usual mudslinging, but Arnold and Alexander conspired to exclude Crockett from the substantial part of the debates. They concentrated upon addressing one another, and studiously ignored Crockett as if his arguments were not worth their attention. The ploy was calculated to convince the voters that Crockett knew nothing of the issues and so could not be considered a serious candidate. Crockett was determined to turn their strategy to his advantage. On one occasion, he spoke first and kept his remarks brief and to the point. Colonel Alexander then raised much the same issues, speaking at length as if Crockett did not exist, and General Arnold in his turn replied only to Alexander's arguments. The general was well into his speech when a flock of guinea fowl came by, clucking and chattering in their manner. The racket they made interrupted the general's argument and unnerved him a little. He asked to have the birds driven off. Crockett let him finish, then approached the stump and in a loud, clear voice congratulated the general for his ability to understand the language of birds:

I told him that he had not had the politeness to name me in his speech, and that when my little friends, the guinea-fowls, had come up and began to holler "Crockett, Crockett, Crockett," he had been ungenerous enough to stop, and drive *them* all away. This raised a universal shout among the people for me, and the general seemed mighty bad plagued. But he got more plagued than this at the polls in August.

(*N*, pp. 204-5)

The incident is a perfect example of the way Crockett's folksy style drew the sympathies of people who were more inclined to identify themselves with the candidate than with the issues. The story as Crockett tells it became one of the most widely repeated yarns in the literature of the Crockett legend.

Crockett's malaria continued to recur, and he was quite ill when he first arrived in Washington. The doctors' bloodletting had again done more damage than the disease. Nonetheless, he went right ahead with his duties, and was hard at work when Congress convened in December 1827. His overriding concern as congressman remained the same elusive goal he had pursued as state legislator: land reform in favor of the West Tennessee settlers. Within three days of the opening of Congress, he had begun promoting his own Tennessee vacant land bill (*DC*, p. 87).

Ironically, the primary reason Crockett never got his land bill through Congress was a virtue of his character that became a liability in the arena of politics. He was the kind of fierce individualist who might very well refuse to do something simply because he was told to do it. He did not like to compromise. He would not trade favors, and congressmen who might have voted with him naturally expected his vote for their pet projects in return. His engaging naiveté and his backwoods pride made him vulnerable to simple deceptions. He was inclined to voice his principles at the most inopportune moments, so that he drove away supporters as fast as he won them. He was irritatingly pushy, insulting colleagues when they did not immediately acquiesce to his demands, as if he did not realize that they did not want to be told what to do, either. The man who so easily won over whole communities by his witty speechifying and story swapping was not especially fitted for the much more complex games of political manipulation. He did not seem to know that common sense is a relative term—that his idea of right action might not be another man's. To win the passage of his land bill, Crockett needed the help of the rest of the Tennessee delegation and of Jackson's supporters at large; but within a short time, he had managed to alienate both groups.

Crockett's campaign in favor of the West Tennessee pioneers was not in line with the aims of the Tennessee political machine as a whole. The state wanted to sell vacant lands at the best prices possible, an intention which was vigorously supported by the landed gentry of Middle Tennessee and by business and professional men whose livings were secure. Actually, the motivation for this position was a worthy one: the state's bill would have provided that proceeds from the sale of open lands would go directly into the establishment of educational institutions. Such was the usual pattern when the federal government turned vacant lands over to a state for disposition. But Crockett's stance was also based on a worthy philosophy. He believed that the children of poor farmers in West Tennessee would never see the inside of a college, and he greatly feared that funding an educational system at the expense of the settlers would open a huge gulf between the two classes, making poor farmers poorer, or even driving them off the land, while giving wealthier folks new opportunities to grow richer still. As James Shackford points out, one crucial factor explaining Crockett's difficulties with his land bill, and his continuing quarrel with Jackson and the Jacksonites as well, was the difference between the interests of the West Tennessee pioneers and the established classes of Middle and East Tennessee (*DC*, pp. 87-137).

In retrospect, the issue of land reform seems a situation ripe with possibilities for compromise and victory. Indeed, a chance to compromise had arisen very early in the game, and Crockett tried at first to take advantage of it. In 1829, James K. Polk, as head of the committee charged with writing Tennessee's official version of a land bill that would represent the interests of the whole state delegation, made some concessions acceptable to Crockett. Crockett did not think Polk's bill was entirely satisfactory, because an amendment gave a pioneer first chance to purchase the land he worked only if he could meet the highest outside bid. Crockett saw the danger in this: it seemed still to classify the pioneers as squatters, and it would have cut most of them out of the action. Nonetheless, Crockett thought he could work with it. He hoped eventually to negotiate a provision saying that a pioneer would have first option to buy at some fixed minimal figure rather than at the level of the highest bid.

This early version of a nearly acceptable bill ran into one special-interest conflict after another and accumulated so many amendments that it was, as Shackford puts it, "talked to death" (*DC*, p. 99). This was precisely the sort of political nonsense that discouraged and irritated Crockett. Ultimately, his acid sarcasm, however justly motivated by a true perception of political hypocrisy, seems to have aggravated his colleagues and thus contributed to the very delays which enraged him. He was caught in a vicious circle.

The land bill ultimately became a reality in one of history's curious twists. In 1839 and 1841, several years after Crockett's death, his son John went to Congress and sought explicitly to secure a bill like the one David had fought for. In February 1841, John succeeded, and the bill was in fact similar to his father's in its most important features. The vacant lands were awarded to Tennessee with the stipulation that the pioneers would have first right to buy at twelve and a half cents an acre (*DC*, p. 239). The early compromise David had made with Polk could have produced a bill like this, but David had not learned—was never to learn—the patience which fortunately did appear in his distinguished son's character.

A major theme of Crockett's *Narrative* is his refusal to be leashed by the Jacksonites. In the very early part of his congressional career, however, Crockett seems to have set aside his suspicions for a while in order to support Jackson. He may have hoped then that Jackson's supporters would back his land bill, but the strongest clue here is his perception of Jackson as a man chosen for leadership by what Crockett calls the "enlightened yeomanry" of the country (see Crockett letter quoted in *DC*, p. 88). Jackson had received a majority of the popular vote for president in 1824, but the Electoral College did not confirm. The House of Representatives subsequently chose John Quincy Adams. Crockett felt that Adams was therefore in office by virtue of the deeds of elitist coalitions and conspiracies, so for a brief time he was able to believe he would be supporting Jackson in the common man's victory that was sure to come in 1828. But Crockett's experiences with the land bill soon led him to think that when the chips were down, the Jacksonites would not favor the enlightened yeomanry over their own elitist coalition, Jackson's entrenched and wealthy backers.

If Crockett was irritated by the general display of congressional politics, with its pork barrel projects, special interests, wasteful spending, and political hypocrisy, then the bold machinations of the Jacksonites literally drove him to distraction, drawing his energies into vociferous complaint and away from steadfast promotion of his own idea of land reform. Crockett simply could not adapt to Washington politics; he kept insisting that games should be played with all the cards visible on the table. He was not the only one burned. Early in the next century, John Quincy Adams's grandson Henry, not an elitist at all, would write in *The Education of Henry Adams*, "The wreck of parties which marked the reign of Andrew Jackson had interfered with many promising careers" (chapter 2).

Crockett's resistance to what he perceived to be a hypocritical betrayal of the common man is exemplified by his opposition to West Point and his abortive fight against Indian removal. In January and February of 1830, Crockett offered several arguments against continuing the United States Military Academy, which had been founded in 1802. All of his arguments can be summarized by saying that he objected to the institution's inherent elitism. He felt that while the academy was supported by a general contribution from all citizens, it was attended only by the sons of the rich. He was not entirely correct in this, for the academy has proved to be an opportunity for quite a few young men who had no money. He felt that officers from the academy would enjoy much exclusive privilege, and in this he was closer to the mark, for very few U.S. Army colonels have ever achieved the rank of general without credentials from West Point. The whole idea struck him as being purely "aristocratic," a "downright invasion of the rights of citizens" (see Crockett's congressional resolution in *DC*, p. 114). And he had not forgotten how he had been ignored when he reconnoitered the Red Sticks and brought his intelligence to Colonel Coffee, who had chosen instead to believe a lost officer who managed to find his way back to camp a day late.

Because Crockett displays considerable anti-Indian feeling in his stories of the Creek War, Professor Shackford and other historians have mistakenly decided that his opposition to the Indian removal bill is inconsistent with his character. Perhaps if the Indians had been "renegades" on the warpath, Crockett's position in 1830 might have reflected his former attitudes; but now he saw Indian removal as an atrocity perpetrated upon neighbors of his, people living in established communities in West Tennessee. No doubt Crockett did have his limitations when it came to social justice, but he was certainly less racist than were his western colleagues, nearly all of whom supported the Indian removal bill enthusiastically.

Crockett's failure to pursue the Indian removal fight openly and vigorously was probably part of an ill-timed attempt to soften the intensity of his style of battle, which by now was doing him visible damage. Such incidents tell us that Crockett's opposition to the Jacksonites, a stance which effectively prevented his ever becoming a successful congressman, was based at least as much in his good intentions as in his naiveté. He spoke out for what he believed in, he refused to go along with slick inside dealers just for the sake of advancing himself, and he never betrayed that enlightened yeomanry of which he was a part. He liked to say that he had remained a Jackson man but Jackson had not. His attitude served his legend far better than it served his immediate purposes in Congress.

MEETING HIMSELF COMING BACK

From 1827 to 1831, Crockett was a member of the Twentieth and Twenty-first Congresses, and he returned to the Twenty-third in 1833. His defeat in 1831 was due partly to his opposition to Jackson, which cost him votes in the more populous and richer areas of his district, but the more critical factor was his substantial loss of flexibility. His constituency, by and large, would have admired his belligerence: as Shackford has said, "The West loved a man 'who refused to sneeze when other men took snuff' " (*DC*, p. 132). But the appalling fact is that for a while this monumental figure in the history of American humor seems to have lost his sense of humor.

His opponent for the 1831 congressional race was William Fitzgerald. Vague allusions in contemporary newspapers indicate that Fitzgerald and Crockett had exchanged vicious insults having to do with either the failure to account for travel allowances (it is not clear which candidate is supposed to have misused the funds) or with Crockett's bad habit of missing congressional roll calls. At one point during the summer of 1831, the two men were scheduled to debate at Paris, Tennessee. Crockett had been loudly declaring his intention to thrash Fitzgerald in public, and Fitzgerald had been forewarned by his friends that this was the day.

When Fitzgerald took the speaker's stand, he laid something covered with a handkerchief on the table in front of him. He began by saying that he had come to prove his side of the quarrel, the charges he had made against Crockett. Crockett rose to shout that he had come to give him back the lie. Shortly, as Fitzgerald's insults grew warmer, Crockett got up and strode toward the stand with fire in his eye. Just before he got close enough to throw a punch, Fitzgerald drew a pistol from beneath the handkerchief and leveled it at Crockett's breast. The embarrassed hero paused, turned around, and found his seat again (*DC*, pp. 132-33). By losing his grip on his own cool style, Crockett had blundered into the wrong role: the master jokester now found himself the butt of a humiliating and dangerous joke. Fitzgerald won the 1831 election by a narrow margin. The subsequent two-year absence from Congress may have helped Crockett regain his composure, at least for the duration of the campaign, for he won the hard-fought contest of 1833.

Little is known of Crockett's activities from 1831 to 1833. He worked hard to expand his ragged farm, he hunted bears and deer, he watched his curious legend growing. The motto "Be always sure you're right—then go ahead" was often quoted when his name appeared in print, and we find him using it along with his signature on business papers as early as 1831 (*DC*, p. 136). In eastern cities and the big river towns, James Kirke Paulding's play *The Lion of the West* had begun to draw tremendous audiences, who were roaring hilariously at the audacious talk and outrageous behavior of Colonel Nimrod Wildfire. Although Paulding denied that he modeled Wildfire after Crockett, there is good evidence that he did so at least in part. Whatever his intention, the public associated the two immediately. The famous character actor, James Hackett, made a strong contribution to this association, playing Wildfire in buckskin breeches and coat and a wildcat-skin hat. Hackett had offered a prize for the best play featuring an American character, and *The Lion of the West* was the winner, so it is probable that his share in fabricating and cultivating the twinship of Crockett and Wildfire was quite deliberate (*DC*, p. 254; see Arpad, "Fight Story" and "Jarvis, Paulding, Wildfire"). Indeed, a drawing of Hackett as Wildfire is commonly taken to be a drawing of Crockett, for it appeared countless times as such in the almanacs and newspapers of the 1830s, 1840s, and 1850s, and again in many books about Crockett published in the twentieth century. It was Paulding and Hackett's Wildfire who made famous the phrase "I'm half horse, half alligator," even though the lines in which the words appear were actually derived from older, widely circulated versions of a formulaic bragging speech supposedly characteristic of the ring-tailed roarers who worked the riverboats and hunted the backwoods. As will be seen in the next chapter of this volume, Mathew St. Clair Clarke loudly proclaimed in *Sketches* (1833) that Crockett himself, not Nimrod Wildfire, had originated the speech, and the absolute fusion of Hackett's Wildfire and Crockett's public image was complete.

In 1832 Jackson was reelected for an unprecedented second term, and the Whigs were naturally in a panic. They had ceased ridiculing Crockett and had begun to exploit his enmity with the Jacksonites even before the end of his second term in Congress, and now they began to cultivate him as a possible candidate for the presidential election of 1836. The Democrats, in turn, stopped trying to deal with Crockett and began making fun of him in speeches, circulars, and articles. Clarke's *Sketches* was rapidly gaining popularity, and the newspapers were busy pirating items from it. Whether the book was designed to lampoon Crockett or to promote him as a potential candidate is not at all clear, but the actual impact it had was the proliferation of the comic components of his legend. Crockett's collaboration with Thomas Chilton on the subsequent *Narrative* was a well-timed response to Clarke's book, serving both to contradict some of the unflattering impressions in Clarke's *Sketches* and to build upon its promotional value. Crockett's repeated references to the possibility that he would consider being a presidential candidate if asked show that he was consciously riding the crest of the fresh surge in his popularity.

The Whigs' exploitation of Crockett worked against itself. It threw into relief the troublesome split that voters observed in Crockett's opposition to the Jacksonites: he seemed to be cultivating his image as a democratic westerner while basking in the praises of the eastern conservatives. The Whigs were not sure of their course. They saw the value of running a man who could both reflect Jackson's frontier style and speak out against the way Jackson had gathered power in the presidency, but they were not prepared to deal with Crockett's cantankerous independence. Crockett's fault lay with his vulnerable ego. Naturally afflicted by that self-consciousness and unnecessary inferiority complex which seems so stubbornly a part of the backwoodsman's character, he responded egotistically to their praises. He loved nothing more than to attend a dinner in his honor.

His famous speaking tour through the eastern states, in 1834, was just the kind of distraction he became addicted to. It was also the kind of activity that caused him to miss roll calls. In Philadelphia, he was presented with a fine rifle. In Baltimore and New York, he was amazed, delighted, and confused by the enthusiasm of the large crowds who turned out to meet him. In March of 1835, a book about his Boston tour appeared; it bore his name as author, but the authors were Whig political writers. (For a discussion of authorship and the issues involved in this book, see *DC*, pp. 184-86.) When it came time to work on legislation, Crockett found that he could not always go along with the Whigs, any more than he could always agree with the Jacksonites. If he felt it necessary to do so, he openly opposed the

Whigs' pet measures. He continued to speak his mind without regard to good manners or protocol. By spring of 1835, it was clear to the Whigs that Crockett would never be a party man, and they abandoned him. In 1836 they would run William Henry Harrison against Jackson's candidate, Martin Van Buren. Both Harrison and Van Buren were consummate party men; the Democrats won with Van Buren in 1836, and the Whigs won with Harrison and Tyler in 1840.

The marvelous ambiguity in Crockett's character and legend is beautifully illustrated by an anecdote widely circulated in newspapers across the country in December of 1833. James Hackett was to appear in a benefit performance at the Washington Theater, and in response to Congressman Crockett's special request, he would perform scenes from Paulding's *The Lion of the West*. Crockett himself was escorted to a reserved seat center front amidst the cheers of a capacity crowd. The curtain rose and Hackett stepped forward, dressed in the resplendent costume of Colonel Nimrod Wildfire. He bowed to Crockett. Crockett rose and bowed to Hackett. The audience went crazy. As he greeted his image, which was greeting him, what thoughts must have crossed Crockett's mind? Perhaps he was disturbed by recognizing that he had begun, all too explicitly, to meet himself coming back. Perhaps, on the other hand, all that mattered was that thundering applause coming down (see Tidwell in Paulding, *Lion*, pp. 7-8; *DC*, pp. 255-56; *Spirit of the Times*, December 21, 1833).

A WORLD OF COUNTRY

In August of 1835, Crockett was defeated in the congressional election, and by early November he was on his way to Texas. He did not purposely go to Texas to fight for its independence, but he did go because he believed that the province would become a new republic, which would mean fresh opportunities for homesteading and politicking. Crockett seems to have left for Texas in a fit of pique after losing the election in Tennessee, and this makes his decision appear to be impulsive. This impression derives from Crockett's own story, written by him in a letter dated August 1835 and widely published later, that he had gotten up in front of his constituents after the election and told them they could go to hell and he would go to Texas. Another version of the story, which was reported by an observer and circulated among the newspapers, suggests that his decision was deliberate, that he had thought of Texas as his alternative even before the election in Tennessee, and that when he arrived in Texas, he began immediately to implement his intentions.

The story was picked up in early April by the New York *Spirit of the Times* a week before it announced the news of Crockett's death, which had occurred in March:

Prentice, the editor of the Louisville Journal, says:—A gentleman from Nacogdoches, in Texas, informs us, that whilst there, he dined in public with Col. Crockett, who had just arrived from Tennessee. The old bear-hunter, on being toasted, made a speech to the Texians, replete with his usual dry humor. He began nearly in this style: "I am told, gentlemen, that when a stranger like myself, arrives among you, the first enquiry is—what brought him here? To satisfy your curiosity at once as to myself, I will tell you all about it. I was for some time a Member of Congress. In my last canvass, I told the people of my District, that, if they saw fit to re-elect me, I would serve them as faithfully as I had done; but if not, they might go to h--l, and I would go to Texas. I was beaten, gentlemen, and here I am." The roar of applause was like a thunder-burst.

(*Spirit*, April 9, 1836)

On January 9, 1835, in San Augustine, Crockett wrote a letter to his oldest daughter, Margaret, and her husband, Wiley Flowers, mentioning his invitations to two upcoming banquets, one in San Augustine and another, probably this one, in Nacogdoches, about thirty-five miles away (see in *DC*, p. 216). The most interesting implication of the newspaper item is provided by its glimpse of Crockett's style at that banquet in Nacogdoches. He is already on stage, the center of attention. He has redesigned his yarn specifically for the delight of the Texans. Clearly, Crockett had begun his Texas electioneering, even though he may not yet have known what office there was to seek.

Crockett was also thinking seriously about the possibility of staking out a very large ranch in Texas. Texas was opening up fast; for some time, agents such as Stephen Austin had been bringing in American settlers, encouraged to do so by the Mexican government itself. Mexico was in turmoil and had been since winning independence from Spain in 1821, and the government did not want to expend money and troops to put down the Indians who controlled the enormous Texas province. A settler and his family could buy over 4,000 acres for a few cents an acre, and, in spite of the risks, thousands of Americans jumped at the chance (see La Fay, pp. 450-54). Like them, Crockett sought a new opportunity for himself and his family, just as he had when he explored the Obion River country. In a letter to Elizabeth's brother-in-law, George Patton, written in October 1835, Crockett says, "I am on the eve of Starting to the Texes . . . we will go through Arkinsaw and I want to explore the Texes well before I return" (see in *DC*, p. 210). He and his party were out on an expedition of discovery. The remote possibility of fighting in some military skirmish had probably not occurred to them at all.

No doubt Crockett expected the survey to be high-spirited adventure, too; but what more practical plan could he have made? A new homestead—a big one this time—would have allowed him once again to wipe the slate clean and start fresh in an open, rich country. Crockett does not seem to have written much to Elizabeth during these last years, and his restlessness may indicate that their marriage had grown cool. On the other hand, negative evidence such as nonexistent letters cannot be construed as a sign of marital strain, and one thing obviously on Crockett's mind was her investment and her largely unrewarded hard work. Elizabeth's well-organized and steadfast management was continually diminished by indifferent crops and burgeoning debts. It is interesting that in 1854, long after Crockett's death and the successful outcome of the rebellion to which he had given his life, some of Crockett's family did what he had in mind all along. Elizabeth, her son George and his wife, and Robert and Rebeckah Crockett, whose parents were David and Elizabeth, accepted an invitation from the grateful government of the new state and settled on the Texas frontier (*DC*, p. 236).

Professor Shackford traces Crockett and the Tennesseans' route down the Mississippi to the mouth of the Arkansas River and up the Arkansas to Little Rock. Then they cut across to the southwest and struck the Red River, probably near Fulton, Arkansas. Following the Red River to the west, they thoroughly explored the country along the northern boundary of Texas. Crockett was enchanted. Had he survived this journey, he probably would have settled in the Red River valley: "It is the pass where the buffalo passes from north to south and back twice a year," he wrote, "and bees and honey plenty. I have a great hope of getting the agency to settle that country and I would be glad to see every friend I have settled thare" (see in *DC*, pp. 214-15). He had found his dream, but instead of lingering here, the party crossed the river and rode south to Clarksville, Texas, and from there to San Augustine. At this point, two of the Tennesseans, Abner Burgin and Lindsey Tinkle, turned back, but William Patton and Crockett signed the Texas oath of allegiance, which would allow them to vote and run for election. By this time, Crockett had discovered a political opening; specifically, he hoped to serve in the impending Constitutional Convention.

Why Crockett then rode on to San Antonio remains something of a mystery. It is possible that he took a stand, not against Mexico, but once more against the Jacksonites. The Americans in Texas had already split into two contending factions, and philosophically the division was more or less along the line between the Jacksonian Democrats and the conservative Whigs. A provisional government was being formed in Texas by the General Council, and this group planned the Constitutional Convention which caught Crockett's notice. But the maverick Sam Houston, who had brought his Jacksonite sympathies with him from Tennessee, had been made commander in chief of the Texan forces by the preceding authority, the Consultation Convention, and he would not relinquish his command. The Alamo was defended by officers sympathetic to the General Council, and they had urged that Houston be replaced immediately. In fact, the Alamo was still manned after January of 1836 only because its defenders did not recognize Houston's authority and so disobeyed his order to withdraw. As if this detail were not ironic enough, the chances are that if Colonel Travis and his officers had obeyed Houston's order, Santa Anna would have bypassed the Alamo and crushed the still disorganized Texan army to the east. The battle of the Alamo weakened Santa Anna's army and gave Houston time to organize his (La Fay, p. 451).

Professor Shackford believes that Crockett's old spite for Andrew Jackson was a strong motivation for his rash decision to go to San Antonio. He further argues that Crockett went to the Alamo to lend support to the young officers who had the nerve to defy Houston (*DC*, pp. 220-21). This seems to assign to Crockett too much cynical deliberation, although he was certainly capable of stubborn single-mindedness. It seems just as likely that Crockett went to the Alamo merely because his dander was up. His adventurous spirit and curiosity would not let him miss what he by now sensed to be some momentous historical turn. His political inclinations helped determine the point at which he would enter the fight, but it was in his nature to fight. For all the confidence expressed in his favorite motto, the truth is that Crockett was a man who would go ahead whether he was sure he was right or not.

Crockett's letter from San Augustine on January 9 to Margaret and Wiley Flowers is the last of his letters to have survived. In it he says that he has been received everywhere with ceremonies of friendship such as those two banquets, and although the trip has been difficult and sometimes dangerous, he is safe, feeling very well, and in high spirits:

> The cannon was fired here on my arrival and I must say as to what I have seen of Texas it is the garden spot of the world. The best land and the best prospects for health I ever saw, and I do believe it is a fortune to any man to come here. There is a world of country here to settle.

He then tells Margaret and Wiley that he has signed up with the volunteers, but urges them not to be concerned about his safety. What he says next reveals Crockett the adventurer feeling well placed in his world at last: "I am rejoiced at my fate. I had rather be in my present situation than to be elected to a seat in Congress for life. I am in hopes of making a fortune yet for myself and family, bad as my prospect has been" (see in *DC*, p. 216). This is not a rationalization. The man knew he did not belong in an office, behind a desk. Nor did he belong behind a plow, though he may not have realized this. He was most at home on his horse, riding west.

THE MANY DEATHS OF CROCKETT

Of all the details of Crockett's biographical history, his death at the Alamo has been most obscured by legend. The picture we have of him swinging his broken rifle and clubbing Mexican soldiers even as they cut him down with bullets and bayonets is a pure invention. Until quite recently, this legendary version of his last moments has simply filled the empty space that was the result of our having no accurate eyewitness account from a survivor on the Texan side. All of the Alamo's defenders died. Two noncombatant survivors, Lieutenant Dickinson's wife, Almeron, and Colonel Travis's slave, Joe, had remained hidden throughout the battle. Their stories have been distorted into the romantic version of Crockett's death. They said they saw him lying dead, horribly mutilated, the bodies of his attackers all about him. They did not see how Crockett actually died. Their reports, which were factual accounts of what they saw after the battle, were taken up and liberally interpreted by the American press, which had been busily fabricating legend out of Crockett's journey even before the fall of the Alamo.

The inevitable book, purportedly based upon a diary kept by Crockett, appeared the following summer. This was *Col. Crockett's Exploits and Adventures in Texas*, "Written by Himself." There was no such diary, and except for the first two chapters, the book was not written by himself. It was written by Richard Penn Smith as part of an agreement between him and the publishing firm of Carey and Hart. The plan was not only to make money on the new book, but to move out all the remainders of that packaged product marketed by the Whigs in 1835, *An Account of Colonel Crockett's Tour to the North and Down East*. Edward L. Carey had observed the sensation caused by news of Crockett's death at the Alamo, and he proposed to Smith that a new book would sell thousands of copies and revive the sales of *Colonel Crockett's Tour* as well.

The first two chapters of *Crockett's Exploits* were based on two letters Crockett had written to Carey and Hart, which Carey had given to Smith. The second of these is the letter which contains the story about going to Texas instead of hell. Smith posed the contents of the two letters as diary entries for July 8 and August 11 of 1835. He then forgot to make the rest of the chapters look like diary entries, until he reached the closing episodes and restored the device in order to pretend that the text came from Crockett's diary at the Alamo for February 19 through March 5. He also forgot to maintain any semblance of Crockett's use of the vernacular. His own text almost immediately lapses into a formal and moralistic hack writer's style which is just the opposite of Crockett's. As if all these machinations were not enough, Smith plagiarized material from other books that must have been lying around the Carey and Hart offices. One of these was Augustus Baldwin Longstreet's *Georgia Scenes*, which had been published in 1835. Smith's third chapter assigns to Crockett an anecdote stolen from Longstreet's book—his famous "Georgia Theatrics," a sketch about a country boy pretending to fight backwoods style with his own shadow.

Smith is said to have written *Crockett's Exploits* overnight, which would seem to be impossible, except that Crockett's letters and the plagiarism from Longstreet (and others, no doubt) helped him get a quick start. He worked over the letters, sent the first two chapters off, and then kept on assembling copy and writing filler just ahead of the printer, so that the book did come out within a couple of days of its inception. Carey and Hart further disguised the book's history by imprinting it with the name of another Philadelphia firm, T. K. and P. G. Collins. The volume contains several unobscured evidences of its actual manufacturer, including a picture of Crockett which bears the imprint of Carey and Hart. The plan worked perfectly; *Crockett's Exploits* and *Colonel Crockett's Tour* were both sold out almost immediately. It is Professor James Shackford who deserves the credit for putting together the details which discount the historical authenticity of *Crockett's Exploits* and which show that *Colonel Crockett's Tour* and another book said to have been written by Crockett in 1835, *The Life of Martin Van Buren* (a mock campaign biography designed to ridicule Jackson's man), were actually Whig party productions (*DC*, pp. 273-81). In spite of, and partly because of, its largely imaginative content, *Crockett's Exploits*, like Clarke's *Sketches*, is a primary source of many of our collective ideas about Crockett's character. Both books belong to the history of the legend rather than to the verifiable history of the man.

Ironically, the spurious nature of *Crockett's Exploits* has caused scholars to discount its report of Crockett's death. Furthermore, the surge of journalistic promotion in the years following the Alamo obscured the report by displacing it with heroic inventions, and the demands of legend making buried it. It was not romantic enough. The book reported that Crockett and five others had surrendered and that Santa Anna had ordered their execution. As it turns out, this account is very nearly correct. Crockett did not surrender willingly, but he did survive the massacre, and he was executed.

Crockett's distinguished biographer, Shackford, died three years before the publication of his definitive book in 1956 (the final manuscript was edited by his brother, John B. Shackford). Since that year, historians have been treated to an astonishing sequence of discoveries which has established a new and historically accurate version of Crockett's death. The key to this picture is a methodically overlooked source which should have been obvious: the Mexican view of the conflict. The long-ignored eyewitness accounts left by Santa Anna's soldiers have finally been brought to light. Chief among these is a vivid, detailed, and now well-authenticated paragraph written by Lieutenant José Enrique de la Peña. His diary was translated by Carmen Perry and published in 1975. Subsequently, this translation and several verifying accounts were examined and interpreted by Dan Kilgore, a past president of the Texas State Historical Association. His splendid reconstruction is presented in a monograph published in 1978 and titled *How Did Davy Die?*

As Mr. Kilgore points out, a few useful Texan reports came out of the

Alamo before the fateful day of March 6, 1836. Commanding Officer William B. Travis wrote that during the initial bombardment Crockett had been seen everywhere in the fort, "animating the men to do their duty" (Kilgore, p. 12). Another witness told of Crockett's "unerring rifle," which "marked down" five Mexican gunners as they stepped up, one by one, to attempt to fire a cannon bearing on the fort. One report told how Crockett barely missed a shot at Santa Anna himself. The general was reconnoitering the fort, being careful to stay out of the range of most men's rifles, but unaware that he stood within range of Crockett's. It was said that Crockett came so close to sniping Santa Anna that the general went into a rage and ordered his army to storm the fort the very next morning. This is a story that deserves to be true even if it is not, for it stands as an emblem of both Crockett's legend and his failures. He always aimed to hit his opponents dead center, and one way to fight back when he missed was to outnumber him by some ridiculously large ratio.

Crockett and five or six other defenders survived the siege and were alive when Santa Anna's troops overwhelmed the Alamo. At about six o'clock in the morning, Mexican soldiers routed Crockett and the others from the various vantage points of their last stand, literally wresting their useless weapons from their hands. Fugitive reports that they begged for their lives are contradicted by the verifiable eyewitness accounts, especially Lieutenant de la Peña's. Kilgore traces the report of their surrender in *Crockett's Exploits* to a letter in the New York *Courier and Enquirer* written from Galveston Bay in June 1836, by a correspondent who had interviewed a Mexican soldier captured in a later battle. Some conflicting historical information makes the report partly unreliable, but the details of the prisoners' last moments are not far off the mark. The *Crockett's Exploits* story says they died heroically, and this news report or one related to it would have been Richard Penn Smith's source (Kilgore, pp. 18-21). Lieutenant de la Peña's highly reliable account tells a similar story of the capture and execution; he says that Crockett and the others behaved well and died bravely.

As soon as the prisoners had been taken, General Manuel Fernández Castrillón of Santa Anna's staff had them brought into the courtyard. In spite of Santa Anna's orders to take no prisoners, Castrillón tried to speak with him in their behalf. He may have known who Crockett was. Santa Anna was angered by this gesture, and ordered the immediate execution of the prisoners. Lieutenant de la Peña records that the commanders and other officers who had been in the heat of the battle "were outraged at this action and did not support the order," but several other officers who had not fought, and who were perhaps interested in flattering Santa Anna, carried out the general's command before anyone could attempt to intervene (Peña, p. 53). The prisoners were bayoneted and shot. The whole incident must have taken no more than a few minutes (Kilgore, p. 47).

Mr. Kilgore tells us that the Texas press and other parties who did not want the legend of Davy Crockett sullied responded with vehement protest to the publication of Carmen Perry's translation and the subsequent dissemination of these facts. Such an uproar is ludicrous but understandable: we always prefer our imaginative reconstructions over embarrassing realities, and since we define our communal identities by our legends, we are naturally offended if they are questioned. But the facts do not really damage Crockett's image, for they are not embarrassing at all. It is unimportant that he did not go down busting the heads of Mexican soldiers. What is important is that he died in the style by which he lived. Lieutenant de la Peña's account ends with this telling observation: "Though tortured before they were killed, these unfortunates died without complaining and without humiliating themselves before their torturers" (Peña, p. 53; Kilgore, pp. 15-17). What more could we ask of either a man or a legend? David Crockett always looked his opponent squarely in the eye, even on that bright and bloody morning when the opponent was death.

2

DAVY CROCKETT— THE FICTIONS

Although a legend derives in some way from actual persons and events, it cannot simply be a factual history. A legend is an ever-expanding series of imaginative tales. A legend is not the same as a myth, although the meanings of the two words overlap. Both exist as stories which embody the most important values of the people who cherish them and keep them alive by telling and retelling them or, in the case of some myths, by living and reliving them. Originally, the perpetuation of legends and myths depended quite literally upon oral retelling and dramatic or ritualistic reliving. Each generation initiated the next. Since the invention of writing, the mass-produced book has to some extent displaced the storyteller, and in the twentieth century, the word "retelling" must also include the preservation and transmission of traditions by means of magazines, recordings, movies, and radio and television programs. However transmitted, "myth" ordinarily refers to a continuously retold story based upon humanity's deepest, most abstract philosophical values, while "legend" refers to a story which has descended from unusual historical events and which reveals the more concrete, consciously supported values of a particular people.

Myths are usually rooted in distinctly religious beliefs, and many have supernatural events as their subjects. These events are symbolic, and the meaning of their symbolism is often a moral imperative, a commandment from on high. A mythic story may or may not have a source in some actual event or person. The stories of Christ and Buddha do have such a source, while the stories of Prometheus and Adam do not refer to individuals, but to the universal recognition of the paradox of consciousness: the human feeling that good and evil do not exist as part of reality objectively considered, but as ideas by which the mind evaluates reality. A myth which takes form as a moral imperative may alter history by motivating the decisions of whole populations. "The Promised Land" and "Manifest Destiny" were two myths of great force which justified and gave meaning to the collective experience of westward-moving Americans. The Indians whose civilization was destroyed by this migration became deeply involved in myths about the evil that had befallen them and the end of their world.

A legend, on the other hand, always has a tangible historical source, and quite often the source can be readily identified. Many legends involve specific persons, like Crockett, whose actual achievements are displaced in the memories of succeeding generations by a rich variety of exaggerated and invented tales. Legends can have supernatural subjects, but these seldom have religious significance: a good example would be an old, persistent ghost story about a unique place—a moor, say, or a castle. The values embodied in a legend are more specific and less abstract than those belonging to myth, and they are always explicit rather than implicit. The Crockett legend could hardly be thought of as being symbolic, because it speaks directly and plainly of self-reliance, the usefulness of humor, and democratic attitudes. Of course, myths can contain legends, and legends can have mythic dimensions.

Both myth and legend depend greatly upon imaginative fabrication by storytellers, which is why the word "myth" is sometimes applied to a certain kind of big lie which is promoted by powerful persons for economic or political gain. Unfortunately, the negative connotations of this usage occasionally spill back into the literary and psychological meanings of the word. In the better sense, myths and legends can be thought of not as facts, but as truthful fictions, because they arise from, and speak directly to, the hopes and fears of the teller and all the persons in the audience. To them, the ideas which inform the stories are truths even if the stories themselves are—as the backwoodsman would say—made up entirely out of the whole cloth. The legend of the Sasquatch speaks to the truth of the mysteriousness of the northwestern rain forest, even to people who are sure that no Sasquatch can exist. The ritual of baptism, one of the most important examples of the living and reliving of a myth, represents an absolutely basic human truth, the universal hope for renewal, rebirth, and everlasting life. Another crucial difference between lying and storytelling is that the liar insists on the literal truth of his lie and consciously deceives people with the intent to do them harm, while the storyteller invites his audience to join him in a contract of voluntarily suspended disbelief for the purpose of celebrating their shared understandings.

David Crockett's history and all the tales about him taken together constitute, not a myth, but an American legend. However, Crockett's legend arose in the context of three myths which held powerful meanings for Americans living at the beginning of the nineteenth century: God's law was to be found in the unspoiled American wilderness; the noble, self-reliant frontiersman was the archetypal American hero; this hero's task was to lead Americans west into that wilderness and toward the fulfillment of their divinely ordained mission. Initially, these myths were consciously employed in the construction of Crockett's legend by professional writers living in the East who understood their appeal. They also understood many of the ironies which attended the imposition of these myths upon real places and real people, and this partly accounts for some of the satirical humor in the legend. The promoters of the legend drew upon Crockett's style and matched his image to the several backwoods traditions which had already been disseminated by journalists and publishers. The audience which avidly consumed the yarns appearing in newspapers, almanacs, and anthologies included both citified easterners who had never seen a forest or prairie, but who indulged freely in the several popular illusions about backwoods life, and rural folks in the West who knew better, but who recognized the humorous, tall language and already thought of themselves as being citizens of the mythic territory.

In trying to understand the deeply human motivations for the behaviors called legend and myth, it is helpful to cite Friedrich Nietzsche's aphorism, "We have art in order not to die of the truth." The truth is that death will come to us all; art is the process of imaginative invention and discovery by which we cultivate our hopes for rebirth and our longings for the triumph of good over evil, feelings which allow us to transcend the limitations imposed upon us by our awareness of our deaths. This is why the rough humor and swaggering egotism in many of the Crockett stories can be directly connected to the suffering inherent in backwoods life, even if most of the stories themselves were taken up and rewritten by people who never lived there. Whenever tall tales or yarns of hilarious misfortune are told or written, teller and listener alike, or writer and reader, imaginatively confront agony, embarrassment, or danger and dismiss its threat in shared good humor. At the same time, not all of the Crockett stories are marked by this sort of grim humor: there are many surprising examples that are touched instead by sentimentality or gentle joking, comic characteristics which also serve as antidotes to the harsh realities of life.

Curiously, except for stories of his death at the Alamo, David Crockett's legend does not spring from the heroic details of actual battles, as the legends of kings and warriors do. Nor did he ever perform feats of strength and intellect so amazing that they became legendary by the very power of their magnitude or originality. In fact, there is so little substance in the biography, and such great wealth in the legend, that Crockett's story serves not only as a fine example of how one legend grew, but as a paradigm for how any legend grows whether or not the facts support it.

More than one process is involved. We can actually observe how a single deed of Crockett's was expanded into a legendary feat simply by being transmitted as a story. If the story comes from his *Narrative*, we can even observe how his own telling contributed to this process. The historical Crockett record exists mostly in writing, and the legendary record exists in writing, pictures, and, more recently, film. Occasionally, there is evidence that some part of a Crockett story had an oral history of its own before it was published with the figure of Davy as its main character. He was also introduced into newly reinvented stories that had nothing to do with his life. His own comic style as a politician and yarn spinner was tall, rough, satirical, and economical; there was already in existence a widespread appreciation of tall, rough, satirical, and economical humor.

Crockett's style of living and storytelling lent itself readily to the adaptation of his image to an existing tradition. Right from the very beginning of the legend-making processes, perhaps as early as his first term in Congress, Crockett became ever more widely recognized as an exemplary tall-talking backwoodsman. The key phrases in Nimrod Wildfire's speech from Paulding's *The Lion of the West* appeared in print at least twenty years before Paulding used them in the play; not long after the play appeared, they were popularly assigned to Crockett. But it would be a mistake to discount the importance of Crockett's imagination to the many complex fictions which constitute the legend. If the literary type of the backwoodsman was an invention, the living backwoodsmen upon whom it was based were known for their inventiveness, and Crockett was known as a very inventive backwoodsman. Crockett fit the type not so much because he imitated it, but because he exemplified it so well. Moreover, the existent formulas were newly re-created by the Crockett who wrote a political autobiography, gave comic speeches, and told humorous stories. Tall yarns already existed, but he was in fact a tall-yarn spinner and the hero of tall yarns. Perhaps we might say that the type and the man were made for each other.

Although scholars not long ago thought of Davy Crockett as a true folk figure appearing in stories which were spontaneously generated and orally perpetuated, it is now clear that the legend as a whole cannot be considered folklore. The question cannot be closed absolutely, because nothing can be proved from the absence of evidence; but so far, all apparent folk forms in the Crockett legend have been shown to be deliberate exploitations. From the beginning, major contributions were made by professional writers, including hacks, who combined the figure of Crockett and the established components of the backwoods tradition in stories designed for publication. If a folklorist should collect a Crockett yarn in some mountain cabin or backwoods hunting camp, chances are the yarn has a printed source. Richard M. Dorson, a folklorist who published a modern edition of some of the Crockett almanac yarns, was able to match a couple of them to folk tales which had other, anonymous heroes. The Crockett legend can thus be thought of as having drawn upon folklore, but it seems unlikely that any of its recorded forms are themselves folk artifacts (see Dorson, *Davy Crockett*, pp. xi-xxvi; and Dorson, *America in Legend*, pp. 64-80).

The key to understanding the growth and dissemination of Crockett's legend lies not in the study of folklore, but in the study of the remarkable adaptations of Crockett's image to existing forms. Furthermore, several features of the legend can be understood by recognizing his own contributions to this process. Crockett created new characteristics for the backwoods type and vigorously exercised the old ones. He was a great yarn spinner, and he had a quick and ready wit. He was conscious of the match between his style and the style of the type, and he exploited this connection to some degree. But his self-indulgence sprang from better motives than the enormous expansions upon his character promulgated by the promoters, reporters, and hacks who were at work on his image before 1830 and whose productions in his name flowed steadily at least until the last Crockett almanac appeared in 1856. Crockett knew what folks liked to talk and joke about, and he understood perfectly how their humor offset the whole range of afflictions which weighed upon their lives. His audience included people who felt threatened by the complexities of authoritative or educated language and by official deceptions or maneuverings; they loved his humor because it deflated politics and politicians, law and lawyers, power and the powerful. The legend makers who succeeded him wrote for this audience and for more affluent people who shared their point of view, and the legend still speaks to precisely the same kind of audience. Crockett's sense of humor really was as large as his legend has portrayed it to be, and we have seen what happened to him when he lost it for a while.

Crockett's legend, then, derives from his style and the way posterity has taken up his style and fit it into the larger context of the American backwoods tradition. Crockett can be thought of as actor, comic character, and storyteller all in one. He was an actor on the political stage, he is a character in the many stories constituting his legend, and he became a storyteller any time he had an audience of one or more and when he wrote his book. As a yarn spinner and politician-actor, Crockett controlled some of the aspects of his legend, but the almanacs and periodicals of the mid-1800s, and other media down through the years to this very day, have expanded it far beyond any of his own intentions. One quality binds all the processes involved in the making of the Crockett legend: the style of the life lived by the backwoodsman, which is inseparable from the style of the yarns told by the backwoods storyteller.

The dominant characteristic of this style is humor, which as a psychological trait supported the backwoodsman's adventurous spirit, his self-reliance, and his irreverence toward authority and institutions. These combinations can be found in much American folklore, and one of the reasons writers in the 1830s were attracted to the emerging figure of Davy is that they perceived him as a folk character; but the style is, of course, easily adapted to the more deliberate means of the technological media, especially the printed page and the illustration. The manipulators of the media, and

Crockett himself, cultivated the figure of Davy as the quintessential example of the backwoods type, who is distinguished by his humorous tall deeds and his humorous tall talk. Which of these two behaviors is the more important is clear: no audience would ever be able to perceive the backwoodsman's deeds as being tall and humorous if it weren't for his talk.

Crockett's legend does not spring primarily from his heroism, but it does retain two heroic dimensions. One of these is secondary to the legend's humor, and the other has a great deal to do with humor. The secondary heroic dimension reflects the actual heroic qualities in the man's character. He was a leader both in the Indian wars and at the Alamo; he was willing to put his career in Congress on the line when it came time to fight all alone against measures he thought unjust; he resolutely searched for new opportunities on the frontier for himself and his family; and he was steadfast and unflinching in the face of death. No amount of historical revision and cynical analysis can mitigate the courage in these actions and attitudes. But his heroic deeds were greatly exaggerated by the newspapers in the months following his death, by the book *Crockett's Exploits* and the many rumors and deliberately falsified accounts it spawned, and by the wholly fabricated Crockett almanacs. Incidentally, false reports that Crockett had survived the Alamo appeared sporadically for many years after he died (*DC*, p. 239; and see Dorson, *America*, p. 80, for a drawing made of Crockett when he was ninety-nine years old!). The heroic qualities of his legend were resurrected and emphasized by the Walt Disney television series and movies of the mid-1950s and in John Wayne's portrayal of Crockett in the 1960 film *The Alamo*.

The comic quality of Crockett's legend, however, is the one that has persisted continuously across time, and this observation suggests the other, primary heroic dimension. There is a certain heroism in humor. As Mark Twain said, it is our best weapon. As a defensive weapon, laughter lets us limit our fears and reduce our worries; as an offensive weapon, it helps us deflate the things which oppress us. The real people living in the real backwoods made humor one of the essential ingredients of their everyday speech. They cultivated irreverence and the satirical, skeptical stance. Like any skillful artist, Crockett was both the master of his tradition and an imaginative contributor to it. All the stories of the Alamo say that he played a borrowed fiddle and told yarns to keep the people's spirits up. Who can measure the value of humor in the face of doom? It must have been even more important than his rifle. Or, since humor is distinctly a weapon, the style of his marksmanship and the style of his humor might productively be compared. It is said that style is the man, and Crockett's sharp-shooting humorous style distinguishes both the man and the legend.

THE STYLE OF THE MAN

Virtually no complete and accurate report of David Crockett's physical appearance is known to exist. In his biography, James Shackford makes a fine educated guess by gathering information from the few portraits still extant and from the reports of Crockett's "honest friends" and "honest enemies." Shackford's use of the adjective "honest" is pertinent: all historians pursuing the facts of Crockett's life have to wade through mountains of deceptive descriptions, some of them well-intended fabrications and some of them lies. Crockett was not six feet four in his stocking feet as the popular phrase puts it; rather, it is almost certain that he was between five feet, eight inches and five feet, ten inches tall (*DC*, p. 282). When he was young, he had a strong, light build, which gave an impression of height. A full-length portrait painted by John Chapman when Crockett was forty-seven shows that by then he had become solidly built—not at all fat, but stocky (see Davis, "Chapman's Crockett"). He stood straight, had a ruddy, open countenance, and looked directly at people when he talked or listened to them. It is likely that he had dark blue eyes and brown hair. Whenever he describes himself during a bout of malaria, he tells how he lost the rosy color in his cheeks (see *DC*, pp. 281-91). Sánchez Navarro, a Mexican officer who saw Crockett when he was captured at the Alamo, called him *anciano*—an oldster. Crockett was not yet fifty, but the stress of battle had taken its toll (Kilgore, p. 38). A few reliable contemporary commentators remarked that they had never seen Crockett in a coonskin cap. Pictures of James Hackett as Nimrod Wildfire depict a hat made from a whole wildcat skin, which Wildfire talks about in the play. Chapman's full-length portrait shows Crockett in a linsey-woolsey hunting shirt, leggings, and mocasins; he is waving a broad-brimmed hat with a rounded crown. With his hair plastered down and his face turned slightly to one side, revealing his arched nose, Crockett looks something like an Indian.

    Contemporary letters and reports commenting upon his death mention not only his heroic behavior, but his "genial warmth," his remarkable ability to be an "agreeable companion" even "amongst strangers," and his "humorous, gentle, human traits" (see letters of Isaac Jones and John Crockett in *DC*, pp. 235-38). A witness at the fall of the Alamo, Señora Candelaria Villanueva, is reported to have said, just before her death in 1899, that she recalled Crockett as "one of the strangest men I ever saw. He had the face of a woman, and his manner was that of a girl. I could never regard him as a hero until I saw him die. He looked grand and terrible, shouting at the front door and fighting a whole column of Mexican Infantry" (*DC*, p. 233). Since this report is second-hand many times over, smacks of masculine journalistic puffery, and fails to match the credible accounts of Crockett's capture and execution, it is almost entirely unreliable. But the details about his face and manner are fascinating, and they pose the question of where the fabricator gathered the impression of Crockett's delicate appearance. If he wanted to inflate the facts to the level of romantic heroism, why would he have invented this particular touch? A small bust portrait of Crockett, painted by John Chapman before he started work on the full-length picture, captures a tender, sympathetic quality in Crockett's countenance. This portrait is in the Alamo today, in the library collection of the Daughters of the Texas Revolution. It supports the congenial recollections of his contemporaries at the time of his death, and it gives us a hint that the journalist who invented Señora Villanueva's deathbed impressions may have had some concrete source for an idea of Crockett's softer features. The remarks of his contemporaries and the Chapman portrait, at least, are good evidence of how people perceived and responded to his gentleness, warmth, and humor. They make it easy to understand why, when news of Crockett's death reached Nashville, grown men and women wept without shame in the streets (see letter of Dr. S. H. Stout, *DC*, p. 238).

All of this information about his appearance and character complements what we know about his personal style of speaking, joking, and storytelling. He had natural, unschooled good manners, and he was softspoken when not riled. These are important traits of character. Comic yarn spinning and friendly joshing are activities that bind people together. They are complex behaviors, and they are constrained by unspoken and often subtle rules. There are times when one talker may interrupt another, and there are times when any interruption would spoil the story's continuity. When a yarn is concluded, another storyteller who wants the floor can tell the first he has told a whopper of a lie, but there would be no degree of seriousness in his accusation. When a challenger opens in this fashion, the group rightfully expects an even bigger whopper from him. Long-windedness is a fatal error. Tellers often depend upon their audiences for responses which advance the yarn—properly timed laughter, for instance, or certain kinds of incredulous questions and genial mockery. As we have seen, many of the characteristics of good yarn spinning, satirical sniping, and joke swapping appeared in Crockett's political speeches. His success in persuading voters seems directly related to his deployment of backwoods humor. His failure in Congress to implement his arguments against the Indian removal bill and for land reform indicates that he did not do well when he set aside humor for invective and moralizing.

John Gadsby Chapman, the celebrated portrait artist who painted the two pictures previously mentioned, also wrote an interesting and valuable account of Crockett's personal style. Chapman describes the sittings, which took place in May and June of 1834, and the discussions the two men had about the content of the full-length portrait, which they had decided upon after the bust had proved so successful. The artist recalls the clothing Crockett wears in the picture, and he remarks upon the rifle he holds, which seemed to Crockett not quite long enough, the large knife in his belt, upon which is inscribed "Crocket" (David thought the second "t" entirely unnecessary), and the three dogs of thoroughly disreputable mixed breed at the hunter's feet.

But even more important than Chapman's observations about the details of the painting are his recollections about Crockett's behavior and speech.

Chapman noted Crockett's good manners immediately. He was both amused and fascinated by Crockett's humorous talk. Without condescension, Chapman praises the clarity, economy, and naturalness of Crockett's language:

> With all the disadvantages consequent upon deficiency in timely educational training, Col. Crockett's command of verbal expression was very remarkable, say what he might his meaning could never be misinterpreted. He expressed opinions, and told his stories, with unhesitating clearness of diction, often embellished with graphic touches of original wit and humor, sparkling and even startling, yet never out of place or obtrusively ostentatious. As for his back-woods slang—it fell upon the ear meaningly and consistent as might the crack of his rifle or his "halloo" from a harricane or from a cane-brake. It was to him truly a mother-tongue, in which his ideas flowed most naturally and found most emphatic and unrestrained utterance.
>
> (Davis, p. 170)

The comparison of Crockett's slang to the crack of a rifle is an illuminating observation upon his style.

Chapman goes on to say that Crockett "rarely, if ever, exhibited either in conversation or manner, attributes of coarseness of character that prevailing popular opinion very unjustly assigned to him." The painter refers to the opinion held by Crockett's political enemies; sympathetic popular opinion as displayed in most newspapers reflects a Crockett who is much more like the man Chapman knew. He records that Crockett never used profanity and that his "narrations of events, and circumstances of his adventurous life," were marked by "an earnestness of truth" (Davis, pp. 171-72; *DC*, pp. 288-89).

It may well be true that Crockett seldom used profanity. The backwoods manner in general reserves vehement swearing for moments of true crisis, when it is both justified and useful. If Crockett did indulge in off-color jokes or remarks, they were doubtless well disguised as double entendres. There is the barest possibility, for instance, that the last word in his favorite expression of surprise, the famous expletive "Wal, I'll be shot!" is a humorous distortion of the commonest Anglo-Saxon excremental verb. It would represent the hypothetical past participle when the verb's simple past is represented as "shat"—a well-known and widely used form. By its very nature, any double entendre is ultimately dependent for its existence upon the listener's or reader's perception, and this interpretation must remain conjectural. Presumably, Chapman noticed nothing off-color in Crockett's talk.

The painter records several instances of Crockett's wit, and two of them are quite revealing. One day, Crockett came to a sitting with a worried look on his face. He had in his hand a crumpled letter. The painting was by now well advanced, and the two men were on good terms, so Chapman felt free to ask what was the matter: " 'I hope you have had no bad news this morning?' " Crockett replied, " 'No . . . 'spose not—only a son of mine out west has been and got converted. Thinks he's off to Paradise on a streak of lightning. Pitches into *me*, pretty considerable. That's all' " (Davis, p. 171).

The other anecdote implicitly gives us a succinct explanation of exactly why Crockett succeeded so well among the voters back home and failed so miserably in Congress. Chapman had met Crockett on Pennsylvania Avenue, and the Colonel, having just come out of an extended congressional debate, looked "very much fagged." Chapman said, " 'You look as tired, Colonel, as if you had just got through a long speech in the House.' " Crockett's answer was,

> "Long speech to thunder . . . there's plenty of 'em up there for that sort of nonsense, without my making a fool of myself, at public expense. I can stand *good nonsense*—rather like it—but *such nonsense* as they are digging at up yonder, it's no use trying to—I'm going home"—
>
> (Davis, p. 173)

The distinction Crockett made between good nonsense and political nonsense constitutes a nicely condensed theory of humor. John Chapman's wonderful record is, of course, biased: we see Crockett in a context which demanded congeniality from both subject and artist. Since he was a quite respectable painter, however, Chapman's eye for detail can be trusted, and his observations upon Crockett's character are reinforced by the other friendly contemporary accounts.

The best direct evidence of Crockett's wit and his facility for storytelling and humor is to be found in the most obvious source, his *Narrative*. Chapter 3 of this volume presents samples from his *Narrative* which illustrate several of his comic techniques: in particular, his use of language formulas representative of backwoods talk, his strong sense of how a good yarn is put together, and his talent for vivid and economical description. He wove jokes and witty remarks into the *Narrative* just as he wove them into his campaign speeches. In both, humorous touches serve as carefully timed counterpoints which provide a control against inflated rhetoric. Humor deflates not only an opponent's pretensions, but any propensity for overseriousness in the humorist's own talk or writing. Wit breaks up the monotony of an extended argument or explanation. In speech or prose, humor surprises and entertains, and any well-timed joke has, at a minimum, one obvious value, which is to recapture the audience's attention.

But Crockett's humor is wider and deeper still: it permeates his whole style. Assimilated into his character was a thoroughly developed and strenuously exercised understanding of humor's great utility. Because it isolates and clarifies pain, humor confirms our humanity; because it is also an antidote to pain, it affirms our humanity as well, by helping us control the meaning of our lives. The skillful humorist is not to be confused with the boor, or lout, who does not know when or how to tell a story. On the contrary, he is one of civilization's few creators of true community. Crockett's rifle-shot wit punched holes in prejudices and broke through barriers to establish a relaxed sympathy between himself and crowds, or among the individuals in any chance gathering which by good luck included him as a member. His was a friendly style, the kind that lets everybody in on the joke so that all might share some brief victory over banality and confusion.

A story from Crockett's *Narrative* illustrates further the readiness of his legendary wit and at the same time reveals the defensively masculine characteristic to be found in some frontier humor. It is a commonplace that one of the appeals the frontier myth has for Americans is that it implies freedom not only from institutional dictates, but from familial demands as well. In general, Crockett was a responsible husband and father; but as mentioned in the preceding chapter, he preferred to go hunting and electioneering while Elizabeth milled the grain at their place on Shoal Creek or managed the homestead in the Obion River country. A married frontiersman felt himself obliged to cultivate his liberty and rule the roost when he was home. This particular story concerns David and Elizabeth's wedding in 1816, which had been planned as a somewhat formal affair. A bona fide minister officiated, guests dressed in their Sunday best gathered in the Patton living room, someone played appropriate tunes on the piano, and the bride was scheduled to enter ceremoniously through the door. In spite of all this elegant preparation, the wedding rite was rudely interrupted just as it began. As everyone turned in breathless anticipation toward the door, a grunting was heard outside, and in trotted a pig. With his usual timing and aplomb, Crockett put the flustered guests at their ease by ushering the intruder out, saying, " 'Old hook, from now on, *I'll* do the grunting around here' " (*DC*, p. 35; *N*, p. 127 and fn. 7). The issue of male pride may be safely set aside; incidents such as this gave Crockett an enviable reputation as a yarn spinner and humorist. His abilities made him a valuable friend to his comrades-in-arms and his hard-pressed backwoods neighbors.

Early in his career as a Tennessee state legislator, Crockett acquired the nickname "the gentleman from the cane." Crockett may have told the following anecdote himself during his conversations with Mathew St. Clair Clarke, or Clarke may have put it together using details he had gathered from Crockett's talk. Either way, the story serves as a good illustration of Crockett's ability to break down barriers, overcome embarrassment, and deflate the opposition's pomposity. James Shackford is not inclined to dismiss the story, and it is he who identifies the opponent, who is known in Clarke's *Sketches* only as Mr. M------l. Crockett had just finished arguing in favor of a measure being considered by the House, when James Mitchell, a finely dressed representative from an older, more populous district, rose to speak against it, referring in a contemptuous way to the previous remarks of "the gentleman from the cane."

Crockett's friends urged him to retaliate, but when he rose to do so, he stammered and hesitated until he had to sit down in mortified silence. Later that afternoon, Crockett returned to the chamber with something concealed beneath his coat. He called for the floor, and when he rose, all the members saw that he had pinned to his shirt a cambric ruffle of exactly the style Mitchell always wore. In his best theatrical manner, he turned slowly from side to side, thrusting out his chest so that everyone could see it plainly. The debate disintegrated in hilarity, and Mitchell, mortified in his turn, hastily fled the chamber. Later, the story goes, Mitchell claimed that he had spoken sincerely, simply intending to say that Crockett was a gentleman who hailed from the canebrake country. The nickname stuck, and whenever people used it, they meant it literally, and its potential as a sarcastic insult was entirely forgotten. Shackford points out that Crockett uses the phrase in the *Narrative* in ways that suggest the story is true (*DC*, pp. 52-53; see *N*, p. 167, fn. 4; Clarke, *Sketches*, ch. 4).

The anecdote, referred to earlier, about Crockett's saying that a premature report of his death was "a whapper of a lie," illustrates his way of taking up a deflationary understatement or witty exaggeration from the vernacular and applying it with expert timing to resolve a difficult situation. Mark Twain's use of the same basic formula reinforces our suspicion that the joke is traditional. Many scholars believe that Crockett's famous aphorism "Be always sure you're right, then go ahead" was likewise already in the language, but that Crockett is responsible for making it a primary feature of our collective memory of the backwoods tradition. According to James Shackford, the story behind the aphorism may be an invention which was originally designed to link Crockett and Andrew Jackson. In December 1813, during the campaign against the Creeks, a young officer went to the general for advice. His men were restless, fatigued, and hungry, and he wanted to know how to keep discipline. He was accompanied, so the story says, by "an awkward, boy-like soldier"—David Crockett. Jackson's advice to the young officer was, " 'Don't you make any orders on your men without maturing them, and then you execute them, no matter what it costs; and that is all I have to say.' " Jackson's condensation of several principles of military leadership was quite economical in its own right, but when Crockett repeated what he had overheard to the other soldiers in his company, he reduced it with style: " 'The old General told the captain to be sure he was right, and then go ahead' " (*DC*, pp. 26-27, and p. 296, fn. 11).

There is no evidence that Crockett invented the maxim, and the story mainly shows how it was widely associated with his name. In the *Narrative*, Crockett made it his for all time in the book's epigraph:

> I leave this rule for others when I'm dead,
> Be always sure you're right—THEN GO AHEAD!

Like Poor Richard, Crockett became famous for the witty exercise of an aphorism he may not have invented. Perhaps partly because the maxim did derive from common speech, it became the touchstone of Crockett's legend, defining in a single shot one of its major values.

HALF HORSE, HALF ALLIGATOR

The most important example of the public's association of Crockett with certain aspects of an already established tradition involves a language formula: the ritualistic bragging employed by two backwoodsmen when they challenged each other to a fight. The definitive example of this particular component of the tradition can be found in an indefinite number of variations, but it always features a braggart who declares that he is "half horse, half alligator" and can "whip his weight in wildcats." In references appearing before 1810, the character making the claim was merely a scruffy frontier renegade; but by the 1820s, he was either called a Kentuckian, the name commonly used for the hunter type, or said to be a keelboatman like the legendary Mike Fink, whose life history is so obscure that Americans still think of him as a fictional character (see Blair and Meine, *Half Horse Half Alligator*). It was in April of 1831 that Nimrod Wildfire, the hero of Paulding's play, first stepped forth upon the stage of the Park Theater and delivered the most famous boast of them all, while the audience howled. In 1833 Mathew St. Clair Clarke attributed the speech to the congressman in his *Sketches*, and thereafter "half horse, half alligator" was most commonly attributed to Crockett, or credited to Paulding along with the assertion that the Wildfire character is a portrait of Crockett.

Actually, Clarke plagiarized the speech from Paulding's play, so the version spoken by Nimrod Wildfire is the central one in the history. A recently discovered copy of the play gives us a text that provides the touchstone for tracing the various traditional phrases. In addition, it was James Hackett's portrayal of Wildfire that provided the twin to Crockett's legendary image, reinforcing the public's idea of Crockett as Wildfire, Wildfire as Crockett, and expanding upon it for a period of several years.

Periodicals spread the speech from Paulding's *The Lion of the West* across the nation and its territories to hundreds of places where the play would never appear. Not long after Hackett had made Nimrod Wildfire famous in New York City, the first issue of a new weekly newspaper was published there; the date was December 10, 1831, and the name of the paper was the *Spirit of the Times*. It was a modest, four-page tabloid, but its ambition was big and broad: the editor, William T. Porter, aimed to cover horse racing, field sports, agriculture, the theater, and literature. To this rich journalistic soup he managed to add reports from the police station, boxing ring, and post office. Like all newspapers of its day, it supplemented its own reporting with an abundance of articles copied straight out of other periodicals. Editors actually encouraged one another to do this by heading good articles "Other papers please copy." The *Spirit* was destined to reign for thirty years as the major medium for the circulation and promotion of American humor, trading its stories with other famous papers like the St. Louis *Reveille* and the New Orleans *Picayune*, and distributing its issues to every town and army outpost in the West, the Far West, and the South. Although the *Spirit* was published in New York, its best writers and most loyal readers lived in the hinterlands, and it would thrive until the Civil War broke its chain of communications. In a prophetic editorial written for the inaugural issue, Porter declared that America would now have a new kind of comic literature: "*Fish* stories, *Wild Cat* and *Panther*, and *Bear* stories."

On the second page of this first issue of the *Spirit*, there appeared, without ascription, byline, or date, the following:

> ADVENTURES OF NIMROD WILDFIRE—"I was ridin' along the Mississippi in my wagon, when I came acrost a feller floatin' down stream settin' in the starn of his boat fast asleep! Well, I hadn't had a fight for ten days—*felt as though I should have to kiver myself up in a salt barrel to keep*—so wolfy about the head and shoulders. So, says I, 'hulloa, stranger! if you don't take keer your boat will run away with you!' So he looked at me slantindickler, and I looked down on him slantindickler—he took out a chor o'tobaccer, and says he, 'I don't valee you tantamount to *that*!' and then the varmint flapped his wings and crow'd like a cock. I riz up, and shook my mane, crooked my neck, and neigh'd like a horse. He run his boat plump head-foremost ashore. I stopt my wagon and sot my triggers. 'Mister,' says he, 'I can whip my weight in wild-cats and ride straight through a crab apple orchard on a flash of lightning. Clear meat-ax disposition—the best man, if I an't, I wish I may be tetotaciously exfluncated!' "
>
> The two belligerents join issue, and the Colonel goes on to say—
>
> "He was a pretty hard colt, but no part of a priming to such a feller as me. *I put it to him mighty droll*—in ten minutes he yelled enough! and swore I was a ripstaver! Says I, '*An't I the yaller flower of the forest!* and I'm all brimstone but the head, and that is aquafortis!' Says he, '*Stranger you're a beauty!* oh, if I only know'd your name, I'd vote for you next election.' Says I, 'My name is Nimrod Wildfire—half horse, half alligator and a touch of the airthquake—that's got the prettiest sister, fastest horse, and ugliest dog in the District, and can outrun, outkick, outjump, knockdown, drag out, and whip any man in all Kaintuck.' "

This item matches, almost word for word, a citation in the *Daily Louisville Public Advertiser* for October 17, 1831 (see Botkin, p. 13). Whoever wrote it down had seen Hackett in Paulding's *The Lion of the West*. He says the play had been given on the previous Friday evening, and in spite of bad weather, the theater was crowded. The editor who selected this item for publication in the *Spirit* would have seen Hackett at the Park Theater, and when he ran across the item in one of his out-of-town exchange papers, he knew it would be recognized by his New York readers and by a great many people who would read it when his paper circulated from town to town along the exchange routes. This is how stories and yarns were distributed far out along the waterways and trails of the West, and back again, in the first half of the nineteenth century. The original reporter of this item may or may not have seen the play in Louisville; it is possible the Louisville editor clipped it from yet another paper, perhaps one from New York (see Arpad, "Fight Story" and "Jarvis, Paulding, Wildfire").

The play opened in New York during April of 1831 and ran until the spring of 1833, when Hackett took it briefly to Washington and then to England for a substantial run in London. Curiously, the script was lost in America, and scholars had only newspaper citations like those from the *Spirit* and the *Advertiser* for evidence of Wildfire's tall talk. In 1951, Professor James N. Tidwell found the play in the British Museum. It had been submitted, as required by law, to the Lord Chamberlain's office for review and approval before being presented at the Theatre Royal Covent Garden (Tidwell in Paulding, *Lion*, p. 10). Comparison of the *Spirit's* report with the play (as it was published in 1954) suggests that the newspaper's transcription is rough; the reporter must have scribbled it down while he was in the theater. In addition, the published play includes revisions made after 1831 by both Paulding and Hackett. The words "exfluncated" and "aquafortis" in the above quotation appear in act 2, scene 2, of the script as the more colorful "exflunctified" and "aky fortis." As in the *Spirit* report, the line "I can whip my weight in wild cats" is here part of the opponent's speech, but Wildfire also uses the phrase earlier in reference to the fighting spirit of his best girl friend. Asked about his hat, he whips it off and displays it with a flourish, declaring that his sweetheart, Patty Snaggs, had made it for him. "At nine year old she shot a bear, and now she can whip her weight in wild cats. There's the skin of one of 'em" (Paulding, *Lion*, p. 35).

However, the phrase "half horse, half alligator" does not appear in this particular scene in the text of the play. It appears in act 1, scene 1, when another character reads a letter from Wildfire announcing his arrival in New York, and the quotation includes yet another formula associated with Crockett, the exclamation that ends in the possible off-color pun discussed earlier. Wildfire's uncle, Mr. Freeman, reads aloud, quoting from the letter:

> "But, uncle, don't forget to tell Aunt Polly that I'm a full team going it on the big figure! And let all the fellers in New York know—I'm half horse, half alligator, a touch of the airth-quake, with a sprinkling of the steamboat! 'If I an't, I wish I may be shot. Heigh! Wake,

snakes, June bugs are coming.' Good bye. Yours to the backbone. Nim Wildfire."

(Paulding, *Lion*, p. 21)

That the printed text here shows Wildfire quoting part of his tall talk indicates Paulding thought of the frontier type as a figure who knew he was a type. The whole characterization, as played by Hackett and as written, confirms this, and Wildfire's self-awareness is part of the parallel with Crockett. Elsewhere in the play, Wildfire shakes hands with his "cordial alligator grip" and several times shouts "I'm a horse!" He argues against the tariff and announces that he will run again for Congress. A shot from his rifle resolves the crisis at the end of the play.

The appearance in the *Spirit of the Times* of a casually recorded version of Wildfire's tall talk was an auspicious event both for the paper and for the history of American humor. The journal rapidly became a most important means by which backwoods expressions, yarns, and lore were collected from, and distributed to, every corner of America. Its thirty-year run constitutes a substantial record of the components of the backwoods tradition, and sometimes the paper's articles indicate where the stories and phrases came from and tell what the public, both in the towns and along the frontier, thought about such things. This vigorous circulation of popular stories and idioms made large audiences conscious of the backwoods tradition and thus helped greatly to promote many legends, including Crockett's. Crockett saw Hackett play *The Lion of the West* in Washington, two years after the *Spirit* was begun, and by then everyone knew Wildfire was the image of Crockett. The *Spirit* reported the climactic encounter of the legendary man and his theatrical counterpart at the Washington Theater in the issue of December 21, 1833.

As for the direct ascription of "half horse, half alligator" to Crockett, however, the single most crucial connection was made in the deliberately perpetrated hoax referred to earlier: in the spurious 1833 biography *Sketches*, Clarke plagiarized several lines directly from Paulding's play, arranged them into a tall speech, and assigned the whole thing to Crockett. *Sketches* is ambiguous in its portrayal of Crockett, and it is difficult to tell from the text what kind of political motives the author might have had. James Shackford established the probable authorship of this book after demonstrating that it could not have been written by James Strange French, in whose name it was copyrighted, and he suggests that the eastern Whigs were behind this venture (*DC*, pp. 262-63).

The Whigs had detected Crockett's shift away from the Jacksonites and had begun to think of him as a possible opponent for a presidential race against Jackson or Jackson's successor. Moreover, Jackson had removed the federal deposits from the Second United States Bank in the belief that its president, Nicholas Biddle, had corrupted its lending practices. (The Independent National Treasury was not established until 1840, when Martin Van Buren decided that the economic depression that had started in 1837 threatened federal funds held in banks.) Crockett favored keeping the deposits in the United States Bank, first as a matter of principle (Jackson had removed the deposits while Congress was out of session); and second as a matter of protecting his connections—the bank had loaned Crockett money, and Biddle had written off the debt (*DC*, p. 171). Shackford says that the probable author, Mathew St. Clair Clarke, was a "literary man," a "historian," and a "friend of Nicholas Biddle" as well as a Whig (*DC*, p. 258). The purposes of the biography would have been to make Crockett as strong a backwoods type as Jackson had been, to portray him as a man opposed to Jacksonian excesses, and to advertise that Crockett had been true to the noble democratic principles Jackson had supposedly betrayed.

The only problem with this possibility is that the book often mocks or lampoons Crockett, even though its anonymous narrator calls for Crockett's reelection to Congress as a first step toward getting him back into national politics. Clarke ought to have been quite interested in helping the Whigs win with Crockett, for he had lost his job as clerk of the House of Representatives in the changeover to Jackson's administration. Like some other Whigs, however, he may have supported the idea of running Crockett, but was not prepared to have his name associated with the ruder aspects of the backwoods tradition. Shackford's opinion, however, is that Clarke could not link his reputation to the book because he was a well-known party man, and his name would have tipped off the public that this was a political biography (*DC*, p. 263). It seems more likely that Clarke's confusion reflected the Whigs' confusion and that the confusions of both were in turn reflected in the book's clumsy handling of Crockett's image. Clarke had spent some time with Crockett in 1828 and, possibly, in 1832, and apparently he remembered quite a few stories that the congressman had told him (*DC*, pp. 119-25). Shackford believes that Crockett may have lent Clarke quite a bit of assistance and later regretted it when he saw that the Jacksonites were cutting the more ridiculous material out of Clarke's book and spreading it around as a caricature (*DC*, p. 258, p. 263).

One Crockett scholar, Joseph J. Arpad, has held out for French's authorship, but his assumptions depend entirely upon the copyright record (Arpad, *A Narrative*, p. 23, fn. 25; Exman, pp. 41-42). The copyright record can be explained by recognizing that Clarke may have gotten some ideas and a little help from several people working on the fringe of national politics, including French. Shackford found evidence that Crockett, Clarke, French, and William Alexander Caruthers traveled to New York together in 1832. He suggests that *Sketches* was copyrighted in French's name as part of the effort to keep Clarke's authorship hidden (*DC*, p. 263). Caruthers, interestingly enough, was author of *The Kentuckian in New York*, an epistolary romance published in 1834, which features a character named

Montgomery Damon who resembles Crockett (*DC*, pp. 263-64; see Arpad, *A Narrative*, p. 34). His title may be yet another piracy at Paulding's expense, for *The Lion of the West* was revised before its production in London in 1833 and retitled *The Kentuckian, or A Trip to New York* (Tidwell in Paulding, *Lion*, p. 9). In 1835, French published *Elkswatawa; Or, The Prophet of the West*, a very bad novel featuring a Crockett-like character named Earthquake (*DC*, p. 258; Arpad, *A Narrative*, p. 34). It is evident that Crockett's disclaimers did not prevent the conspirators from continuing their game for at least two or three more years.

The possibility that Clarke and his friends did spend some time with Crockett in 1832 suggests that *Sketches* may have been one of at least three books resulting from a casual conspiracy motivated by literary aspirations, political maneuvering, and a simple penchant for the hoax. The hoax could hardly be called a rare phenomenon in American literary history, and it may be the common denominator of the entire body of Crockett literature. Perhaps the conspirators simply saw an opportunity to attach their fortunes to Crockett's rising star. At first, in 1832, he may have encouraged them. In one way or another, Crockett inadvertently contributed to some important artifacts of his legend by talking to Clarke and his friends. The most important of these artifacts is of course Clarke's *Sketches*, which can be thought of as the first major Crockett book and as the outrage which motivated Crockett to write his own book.

All of this peculiar history illuminates the processes by which "half horse, half alligator" came to be a popular description of Crockett's legendary character. After stating that the colonel must be made to run again for Congress, the narrator of Clarke's *Sketches* chooses to represent him in "one of his quirky humours." The wording is a good example of the book's mixed rhetoric. Posed as a quotation from Crockett's own storytelling, the crucial plagiarism begins:

> "I had taken old Betsy," said he, "and straggled off to the banks of the Mississippi River; and meeting with no game, I didn't like it. I felt mighty wolfish about the head and ears, and thought I would spile if I wasn't kivured up in salt, for I hadn't had a fight in ten days."

The Crockett character goes on to say he challenged a boatman coming down the river, and the other looked at him "slantendicler," and Crockett looked back at him "slantendicler." This reduction represents a loss from Paulding's text, which reads, "So he looked up at me 'slantindickular,' and I looked down on him 'slanchwise' " (Paulding, *Lion*, p. 54). The Crockett of Clarke's *Sketches* then recalls that he said

> " 'Ain't I the yaller flower of the forest? And I am all brimstone but the head and ears, and that's aqua-fortis.' Said he, 'Stranger, you are a

beauty: and if I know'd your name I'd vote for you next election.' Said I, 'I'm that same David Crockett. You know what I'm made of. I've got the closest shootin' rifle, the best 'coon dog, the biggest ticlur, and the ruffest racking horse in the district. I can kill more lickur, fool more varmints, and cool out more men than any man you can find in all Kentucky.' Said he, 'Good mornin', stranger—I'm satisfied.' Said I, 'Good mornin', sir; I feel much better since our meeting'; but after I got away a piece, I said, 'Hello, friend, don't forget that vote.' "

(Clarke, *Sketches*, ch. 11)

The parting shot seems to indicate political support for Crockett, while the remarks about his belligerence seem to belittle him.

The mixed and occasionally puzzling diction in the above quotations suggests that the author of *Sketches* was naive about some of the idioms he introduced into the plagiarized lines. Joseph Arpad believes that "ticlur" might have been an obscene pun, and that "kill more lickur" would have been an insult ("Fight Story," p. 158). Actually, a tickler was a hunting knife, and backwoodsmen were proud of being able to kill liquor, which meant drinking steadily without getting drunk. If these two images represent a phallic pun and an insult, the meanings are not likely to have been perceived by the author of the book. Crockett may have bragged to Clarke about his drinking, but in his *Narrative* he talks about the moderate use of whiskey as if it were a commonplace remedy. He might have made the "ticlur" pun in conversation, but he would not have permitted it in print because he would have been aware of its second meaning. These usages are blunders which may tell us much about why Crockett denied having anything to do with Clarke's book after it was published.

Two chapters later, Clarke's *Sketches* tells about Crockett's first journey to Washington. This is supposed to have happened in the winter of 1827, when he first took office as congressman. Along the way, Crockett stops in taverns and entertains the folks with his yarn spinning and tall talk:

" 'I'm that same David Crockett, fresh from the backwoods, half-horse, half-alligator, a little touched with the snapping-turtle; can wade the Mississippi, leap the Ohio, ride upon a streak of lightning, and slip without a scratch down a honey locust; can whip my weight in wild cats,—and if any gentleman pleases, for a ten dollar bill, he may throw in a panther,—hug a bear too close for comfort, and eat any man opposed to Jackson.' "

(Clarke, *Sketches*, ch. 13)

Here, the claim that he can whip his weight in wildcats has been taken away from the opponent in Wildfire's scene as Paulding wrote it and assigned to the congressman. Now the touchstones of the fight challenge have been pulled together in a single Crockett speech.

Later, Clarke dutifully reports that Crockett remained true to the earliest of Jackson's principles, but that Jackson himself had turned around and Davy had to oppose him. In subsequently writing his autobiography, Crockett tried to revise his history by making his opposition to Jackson appear consistent. Another thing he did to offset the impressions left by Clarke's *Sketches* was to avoid all semblance of Nimrod Wildfire's tall talk. But by early 1834, the association between Crockett and Wildfire had been well established, and the accounts of Davy's speech and manner written thereafter nearly always say that he was one of the half horse, half alligator breed of men, and, thanks to Mathew St. Clair Clarke, that he could whip his weight in wildcats as well.

The key phrase in the fight speech can be found in a song by Samuel Woodworth, who was an editor of the New York *Mirror* and a playwright as well as a songwriter. His best-known song is "The Old Oaken Bucket." Equally famous in its own time was "Hunters of Kentucky," which was first sung by the entertainer Noah Ludlow to an audience in New Orleans in 1822. The setting was appropriate, for the lyrics celebrate the slaughter of the British soldiers who marched in ranks against General Jackson's militia in the Battle of New Orleans. The Americans were hiding in ambush, and they were armed with their own Kentucky rifles. None of the redcoats reached the barricades, and fewer than ten militiamen were killed. One stanza of Woodworth's song says that the hunters raised a breastwork, or bank, not so that they could hide behind it, but so they could use it as a rest for their rifles. The stanza concludes:

> Behind it stood our little force—
> None wished it to be greater,
> For every man was half a horse,
> And half an alligator.
>
> (Botkin, pp. 9-12)

Given the lopsided outcome of the battle, the boast cannot be judged idle.

The song firmly established a connection between "half horse, half alligator" and the popular image of the Kentuckian. The further association, in the 1830s, of Crockett, a Tennessean, with the legendary figure of the Kentuckian is not a mystery. As mentioned earlier, revisions of Paulding's *The Lion of the West* were titled *The Kentuckian*, and the states of Kentucky and Tennessee were strongly linked in the mind of the public who saw the play in eastern cities. The confusion of Crockett with Daniel Boone, who explored Kentucky and opened the Ohio Valley to settlement, was common in the middle of the nineteenth century and persists today. Jackson's militiamen at New Orleans came from Tennessee as well as Kentucky. The fact that Crockett had help with his *Narrative* from a Kentuckian, Thomas Chilton, is fortunate: between the two of them, they had an extensive knowledge of the backwoods vernacular, and they probably also understood what city folks and easterners expected it to look like in print. Kentucky and Tennessee were in fact related historically, and in the minds of the public, they were the same mythic territory.

Even earlier literary usages of "half horse, half alligator" appear in Washington Irving's *A History of New York*, by "Diedrich Knickerbocker," and a collection of letters written by a traveler, Christian Schultz, Jr., who published his observations as *Travels on an Inland Voyage through the States of New-York, Pennsylvania, Virginia, Ohio, Kentucky and Tennessee, and through the territories of Indiana, Louisiana, Mississippi and New-Orleans*. There is more to Schultz's title, including the dates 1807 and 1808, but the list of states and territories is worth repeating because it indicates the spread of the region to which his observations apply. This was the West in the first decade of the nineteenth century. The citation from the Knickerbocker *History* is dated 1809. Schultz's book was published in 1810; the letter containing the report is dated April 13, 1808, and was written from Baton Rouge in the West Florida territory.

Washington Irving's famous mock history satirizes both historical persons in New Netherland and the style of the historians who had written about them. In one episode, Diedrich Knickerbocker describes a "vagabond" who is skulking around a garrison manned by enemies of "Peter the Headstrong"—a caricature of Peter Stuyvesant. The vagabond is Gallows Dirk, a half-breed outlaw, who wears clothes that mix the styles of the frontiersman and the Indian, and who is equipped with a rusty knife and an ancient fowling piece. He is one of those vagrant "beings" who "infest the skirts of society like poachers and interlopers," his life is "a kind of enigma," and his existence is "without motive." He "comes from the Lord knows where" and "lives the Lord knows how, and seems to be made for no other earthly purpose but to keep up the ancient and honorable order of idleness." He displays a "sharking demeanor" and overhears things while slyly looking for advantages. Since this itinerant free-lance scout acknowledges no allegiances and has a loose tongue, he later sells information to Peter, who is thus enabled to foil his enemies' plot against him. Irving compares the half-breed outlaw to the backwoodsman of Kentucky in a most unflattering manner:

> It is an old remark that persons of Indian mixture are half civilized, half savage and half devil—a third half being expressly provided for their particular convenience. It is for similar reasons, and probably with equal truth, that the backwoodsmen of Kentucky are styled half man, half horse and half alligator by the settlers on the Mississippi, and held accordingly in great respect and abhorrence.
> (Irving, *History*, p. 254)

In telling how Dirk manages to survive several days of severe traveling in the woods before he gets to Peter with the news, Knickerbocker grudgingly admits that the outlaw has endured "a world of hardships that would have killed any other being but an Indian, a backwoodsman or the devil" (p. 255).

Christian Schultz writes that he overheard his version of the fight challenge on the Mississippi levee at Natchez:

> The evening preceding that of my departure from Natchez being beautiful and bright, I walked down to the Levee, in order to give some directions to my boatmen. In passing two boats next to mine, I heard some very warm words; which my men informed me proceeded from some drunken sailors, who had a dispute respecting a *Choctaw lady*. Although I might fill half a dozen pages with the curious slang made use of on this occasion, yet I prefer selecting a few of the most brilliant expressions by way of sample. One said, "I am a man; I am a horse; I am a team. I can whip any man *in all Kentucky*, by G-d." The other replied, "I am an alligator; half man, half horse; can whip any *on the Mississippi*, by G-d." The first one again, "I am a man; have the best horse, best dog, best gun, and handsomest wife in all Kentucky, by G-d." The other, "I am a Mississippi snapping turtle: have bear's claws, alligator's teeth, and the devil's tail; can whip *any man*, by G-d."
>
> (Schultz, *Travels*, vol. 2, pp. 145-46)

It is too bad Schultz did not choose to record more of what he heard, for another "half a dozen pages" of this "curious slang" would be quite useful to us now.

The modern editor of Schultz's book, Thomas D. Clark, writes that he believes the letter from Natchez-under-the-Hill might be the earliest source of "half horse, half alligator" (p. xxiv). Joseph Arpad traces it in Paulding's writings as far back as 1817 and from there to Schultz's *Travels*, which Paulding had almost certainly seen. Arpad also shows how Paulding's friend, the painter and famous tall-tale telling raconteur John Wesley Jarvis, made some contributions to the characterization of Wildfire during the composition of *The Lion of the West*. Arpad quite rightly points out that the sources for "half horse, half alligator" are not in folklore, but in literature, and he suggests that an even earlier source than Schultz's might be found (see Arpad, "Fight Story" and "Jarvis, Paulding, Wildfire").

Washington Irving and Christian Schultz portrayed the backwoodsman as an unlettered and brutish outcast who roamed the frontier because civilization would never have tolerated his presence. Fortunately, by the time the legendary Davy began to talk tall, the backwoodsman's image was no longer quite so dangerous, dirty, or despicable. The writers and politicians who promoted the association of tall talk with the legendary figure of

Crockett often cleaned up the talk and the traditional image as well. Those not so friendly to Crockett did not mind extending the idea of the backwoodsman as an ignorant lout, however, and there are strong traces of this seamier dimension of the tradition in Clarke's *Sketches* and in the Crockett almanacs. On the whole, however, the public's idea of Davy in the 1830s was that he was a hero who possessed naturally noble qualities and had not lost his backwoods sense of humor.

The "half horse, half alligator" formula has persisted both as a component of the Crockett legend and as a component of the more general backwoods tradition. The *Spirit of the Times* for March 27, 1841, featured Thomas Bangs Thorpe's famous story "The Big Bear of Arkansas," whose narrator is a literary man traveling on a steamboat and observing the other passengers. Among those who attract his attention is a " 'plentiful sprinkling' of the half-horse and half-alligator species of men, who are peculiar to 'old Mississippi,' and who appear to gain a livelihood simply by going up and down the river." The hero of the story is Jim Doggett, a backwoodsman from Arkansas, who tells a yarn about pursuing an unhuntable bear, the biggest in all creation, and who calls himself the Big Bear of Arkansas as well. Thorpe's characterization shows that he was thoroughly familiar with the legendary image of Crockett and deliberately adapted it to his backwoods hero. (The story can be found in Cohen and Dillingham, *Humor of the Old Southwest*; see also the article by Lemay.)

Half a century later, Mark Twain made good use of the tradition, especially in the keelboat scene from *Life on the Mississippi*, which is repeated in some editions of *Huckleberry Finn*. In this, one braggart says he is "the old original iron-jawed, brass-mounted, copper-bellied corpse-maker from the wilds of Arkansaw!" He takes "nineteen alligators and a bar'l of whiskey" for breakfast when he's healthy, and "a bushel of rattlesnakes and a dead body" when he's ailing. He concludes, "I'm the bloodiest son of a wildcat that lives!" The other is "the pet child of calamity," and he declares, "when I'm thirsty I reach up and suck a cloud dry like a sponge; when I range the earth hungry, famine follows in my tracks!" Both braggarts are soundly thrashed by a small, wiry "black-whiskered chap" named Little Davy who makes them "own up that they was sneaks and cowards" (a text and discussion of the textual problem can be found in the Norton edition, which is cited in the bibliography).

The line "I'm half horse, half alligator" has even been repeated quite recently in several motion pictures containing braggart scenes. Ernest Borgnine uses it while trying to torment Spencer Tracy into a fight in *Bad Day at Black Rock*. Crazy characters make comic boasts which include the line in *The Life and Times of Judge Roy Bean* and *Jeremiah Johnson*. An astonishing takeoff occurred on television during the 1979-1980 season when an alligator on *The Muppet Show* popped up and said, "I'm half horse, half watermelon!" The whole hilarious event took less than three seconds. Crockett would have loved it.

It is probably safe to say that the "half horse, half alligator" formula goes back further than 1808. As every reader of *Beowulf* can see, analogous speeches appear in the earliest literature written in English, and as the reader of *The Iliad* can testify, heroes are braggarts in the earliest stories of western civilization. Brags and challenges are steps in the hero's progress toward his triumph and his fate. Unmet challenges and bragging that is not matched by performance are often the marks of the hero's antagonist. But tall talk is a mock-heroic form, and the more immediate predecessors of the characters who use it are to be found in the seventeenth-century *commedia dell'arte* in the stock character of the braggart soldier, and in a famous collection of tall yarns, many of them based upon ancient folk tales, told by the persona of *Singular Travels, Campaigns, and Adventures of Baron Munchausen*, which first appeared in 1785.

The Baron was real: he was a German nobleman who had served as a Russian soldier and was famous as a braggart. But the comic persona was the literary invention of Rudolph Erich Raspe, who stole Munchausen's notorious name and attached it to a collection containing hundreds of old European yarns. They were published in English as a sequential narrative. Later editions expanded the collection enormously, and after Raspe's death, the game became a public property, proliferating in scores of spurious derivations and growing ever more outrageous. Hundreds of these ancient tall yarns found their way into American periodicals and anthologies, and even Davy Crockett has from time to time been compared to Baron Munchausen.

The braggart soldier from the *commedia dell'arte* often had the name of Capitano Spavento della Valle Inferna, which is freely translated from the Italian by Walter Blair and Hamlin Hill in *America's Humor* (1978) as "Captain Fear of Hell's Gulch." Other names assumed by various players at various times in the sixteenth and seventeenth centuries included Captain Crocodile, Captain Rhinoceros, Captain Lion-Tamer, Captain Black-Ass, and Captain Earthquake. Traveling companies made later forms of *commedia dell'arte* quite popular in American theaters in the early nineteenth century. The bragging soldier's names and his style may be clues to an earlier source for the "half horse, half alligator" type of man. Blair and Hill, at least, see fit to make a strong comparison between Davy Crockett and Captain Fear (p. 134).

Tracing these analogies shows that a tradition which is an important part of the Crockett legend has its roots in other places and other times, some of them seemingly remote from nineteenth-century Tennessee or Texas. The later echoes show that the tradition continues, sometimes in forms tangential to the Crockett legend. Whatever its ultimate origin, tall talk proved distinctly suitable to the expansion of comic lore based on Crockett's image, even as Crockett's image provided a means for the unbounded expansion of the tradition. Such connections suggest that it would not be wise to cling to the notion that all the dimensions of the legend are purely American. Furthermore, the international reception of the legend's artifacts, from the great success of James Hackett's Nimrod Wildfire in England to the popularity of Walt Disney's Davy Crockett in Europe and Japan, shows that it has a universal appeal, which in turn suggests that it contains many universal dimensions.

THE ALMANACS

The Crockett almanacs are part of a well-established tradition which in America dates back to 1639 and the first New England almanacs. Franklin Meine, who collected the important Nashville series of Crockett almanacs (1835-1838), pointed out that the comic material which dominates the Crockett almanacs also represents a tradition, because humorous touches appeared in the New England almanacs as early as the 1680s (pp. xiv-xv). Benjamin Franklin's *Poor Richard's Almanack* appeared from 1733 to 1758 and, like most almanacs then and since, drew its information from many disparate sources. Very few, if any, of Poor Richard's famous aphorisms are original with Franklin; some of them have analogues in classical antiquity. *The Old Farmer's Almanac* began in 1793 and managed to survive the piracy and casualness which afflicted the others, and derivations continue today. The offhand, free-spirited, and eclectic borrowing which characterizes the tradition constitutes a sort of almanac maker's license. The Crockett almanacs began with an especially audacious exercise of this license.

Like all of the earlier kinds, the Crockett almanacs contain agricultural information, weather predictions, and practical domestic hints. Where the others offer descriptions of farm animals and farming techniques, the Crockett almanacs concentrate on wild animals and hunting yarns. They were printed as small pamphlets of three or four dozen pages on lightweight paper; consequently, they are now extremely rare. There is a great variety in the wording and typography of the titles. Included are *Davy Crockett's Almanack, of Wild Sports of the West, and Life in the Backwoods*; *Crockett's Yaller Flower Almanac*; *Ben Hardin's Crockett Almanac*; *Crockett Comic Almanac*; and finally, *Crockett Almanac*. The earlier ones are illustrated with crude woodcuts of comic scenes; later issues show more sophistication, but the scenes become more violent and even repulsive. Meine's edition of the Nashville series (1955) is quite useful to the study of the stories and tall talk which are associated with Crockett, as is Richard M. Dorson's *Davy Crockett, American Comic Legend* (1939), a collection based entirely on the almanacs. Constance Rourke's *Davy Crockett* (1934) includes a descriptive bibliography which numbers forty-four separate issues running from 1835 to 1856. (Information about the locations of Crockett almanacs and new bibliographical research can be found in chapter 4 of this volume.)

If there can be such a thing as a folk literature which has its primary existence in published form, the Crockett almanacs would seem to qualify. They contain a great many backwoods yarns, a few of them from the oral tradition; they record home-grown agricultural lore and folk wisdom about wild creatures and weather; their illustrations, especially the earlier woodcuts, look spontaneous and unschooled. But the truth is that the Crockett almanacs, like Clarke's *Sketches* and Smith's *Crockett's Exploits*, began with a professional writer's fabrication—a hoax, an exploitation, a literary invention. And while many of the comic yarns have frontier subjects and are written in a backwoods style, the history of the almanacs does not begin in the West, and it has more to do with printing technology than with oral storytelling. Apparently, the history of how the almanacs began is more closely related to the history of their New England predecessors, for the Nashville series seems to have been conceived, assembled, and published in Boston. As Professor John Seelye has put it, they are one of the many examples of eastern-bred western fictions whose existence suggests that "our premier mythic terrain, the trans-Mississippi West, was from the beginning a creation of the East, being almost entirely a territory of the literary imagination" (Seelye, p. 110).

None of this reduces the value of the Crockett almanacs a whit. Instead, it reinforces the idea that the Crockett legend is a set of popular fictions deliberately designed to appeal to basic values held dear, in spite of the facts, by American readers. Seelye's marvelous detective work was published in 1980 as an essay appropriately titled "The Well-Wrought Crockett: Or, How the Fakelorists Passed through the Credibility Gap and Discovered Kentucky." The subtitle draws upon the long-standing public confusion of Crockett and Daniel Boone, who led the settlers through the Cumberland Gap and into Kentucky. In the essay, Seelye displays what he admits to be circumstantial evidence, but it is the kind of circumstantial evidence that holds up in court. James Shackford and Joseph Arpad had already suggested that the almanacs might be a hoax, largely because so much of the material in them comes from Clarke's *Sketches*, Crockett's *Narrative*, and Smith's *Crockett's Exploits*.

Seelye begins by trying to solve the puzzle of a character who appears in the second Nashville almanac, published for 1836. The character's name is Ben Harding (occasionally "Hardin," as in the title of an 1842 almanac), and in 1839 he also becomes the supposed editor of the almanac and the holder of its supposed copyright. The first issue, published in 1834 for the year 1835, pretends to be copyrighted in the legendary name of Davy Crockett. There was a historical Ben Hardin (not Harding), a congressman from Bardstown, Kentucky, who was quite a colorful fellow and a humorist, but he does not seem to resemble the almanac character. In fact, the Ben Harding character in the almanacs is a Kentucky congressman only initially, for very soon he becomes a nautical stereotype, a rough, practical-joking sailor who threatens to crowd the figure of Crockett out of the almanacs. By 1841, however, the perpetrator of the almanacs had recognized the value of keeping Crockett out in front, and Ben, as Seelye writes, "had grown a wooden leg and established contact once again with Davy Crockett, who was alive and as well as could be expected while working in the Mexican mines" (p. 96). The perpetrator drew upon, and in turn promoted, the rumors that Crockett had survived the Alamo.

Why, asks Seelye, was a Kentucky congressman transformed into a saltwater sailor, and what was he doing on the Mississippi River? By an ingenious technique of matching various illustrations from the Crockett almanacs to those from other publications, Seelye establishes that the Nashville series was most likely assembled and produced by Charles Ellms, a Boston illustrator and editor. A picture in the almanac for 1839, titled "The Pirates Head" and supposedly drawn by Ben Harding, exactly matches one titled "the head of Benevides stuck on a pole" in *The Pirates Own Book*, published in 1837 by Samuel N. Dickinson in Boston. This book is credited to Ellms, who later wrote sensational seafaring stories illustrated with woodcuts like those in the Nashville almanacs.

Ellms also started two other comic almanacs in the early 1830s, the *American Comic Almanac* and the *People's Almanac*, and he published these for several years before turning them over to the Dickinson firm. The first issue of the *People's Almanac*, printed in 1833, contains a woodcut depicting a fight between a ferocious alligator and an enormous snake. On page 17 of the Crockett almanac for 1836, there appears an illustration titled "An Alligator Choked to Death." It shows how Ben Harding, in this issue still a Kentucky congressman, kills an alligator: he dives down its throat and lodges therein, choking it to death. The two alligators are one and the same. The snake that appears in the *People's Almanac* version has been deleted for the Crockett almanac, and Harding's legs have been drawn in, protruding from the alligator's open jaws. All the other details match exactly, right down to the inclusion of a baby alligator which can be seen hatching from an egg beneath its mama's left rear leg. The correspondences between these illustrations suggest, in Seelye's view, "a certain printshop propinquity" (p. 103).

None of the letters that provide the evidence establishing Ellms's connection with the woodcuts used in the Nashville almanacs is dated later than 1834 (Seelye, p. 105), so there remains the possibility that the original blocks were somehow transferred from Boston to Nashville. That a woodcut artist in Nashville copied pictures from Ellms's publications with a thoroughness comparable to present-day photocopying is, of course, out of the question. But if the blocks were transferred to Nashville, the fact would not significantly alter Seelye's main point. The almanacs still could not be considered folklore in the usual sense. They would simply become a western invention rather than an eastern invention, and they would remain a product of someone's "literary imagination," a hoax based largely upon previously printed sources. By 1841, the history of the almanacs is definitely centered in New York, although a few were published in Boston later than that and stray imitations appeared sporadically in various other towns (see Rourke's descriptive list, *Davy*, pp. 252-58).

Material from the earlier Crockett books kept the almanacs going for quite a few years, and the early, relatively artless woodcuts give them their folksy appearance. At first, the almanacs do not seem to be politically motivated. By 1843, however, when they were taken over by the New York publishing firm of Turner and Fisher, their contents become slicker, more sophisticated, and distinctly political. As Seelye writes, they "begin to express a spirit that is closer to the dark side of Jacksonianism than to the Whig élan." The grotesque illustrations grow ever more violent. The language of latent racism displayed in the early issues erupts into an inflated, genocidal jingoism. American expansionism is depicted as a justified displacement of degenerate colored races (Seelye, p. 97). On the one hand, the later almanacs can be given credit for extending the Crockett legend; on the other, we may wonder how the legend managed to survive them.

Considered separately from the almanacs' various political and social distortions, the sheer variety of the yarns appearing in the whole twenty-year run is astonishing. The range marks the extremities of the Crockett legend. Some of the anecdotes are racist, crude, and violent, and a few are even grossly excremental. At the other end of the scale, some are downright mythic, attributing to Crockett the qualities of gods and heroes—Apollo, Hercules, and Prometheus.

The absolute low point of the Crockett legend, the worst almanac yarn in the whole lot, is a grotesque story about a squatter who wants to prove to the land agent that he has lived on a homestead and begun to develop it. Such proof would allow him to buy the land at the nominal price. He asks Davy to appear on his behalf and swear that he had observed the squatter making improvements three or four years earlier. But Davy knows the man was out of the country at that time, so he challenges his veracity. The squatter takes Davy out to the clearing, where he has built a shack and installed some animals. There he tries to trick Davy into believing that he has established a claim, but the trick is too transparent. Fiercely insulted, Davy points his rifle at the neighbor and forces him literally to eat his punishment—with a spoon: " 'Do you see what that cow has just let drop? It ar not honey or apple sarse, ar it? Now if you don't sit down and eat every atom of it, I'll make daylite shine through you quicker than it would take lightning to run round a potato patch' " (*The Crockett Almanac*, 1839, pp. 11-12; or see Dorson, *Davy Crockett*, pp. 89-90). The squatter complies, and later he apologizes, properly mortified.

The mythic dimensions of the almanac yarns are nicely displayed in Meine's collection, and elsewhere in Dorson's, and they form the basis for Constance Rourke's resonant narrative of the legendary Crockett. Davy steered an alligator up the Niagara Falls, escaped from a tornado by riding a streak of lightning, climbed a mountain to wring the tail off the comet, swallowed a thunderbolt, and drank up the Gulf of Mexico. When he was eight years old, he weighed two hundred pounds. His girl friend was fat as a bale of cotton, and when she sneezed, leaves fell from the trees for miles around. His youngest daughter wrestled bears, and she had apparently inherited her father's famous grin: where he had once grinned the bark off a tree, she grinned a pack of wolves into "total terrifications." He killed four wolves when he was only six, shaved a panther, and bit a rattlesnake to death. This is the Davy who, when the whole planet froze over, thawed out the earth's axis with bear oil and kicked it loose again. When the sun came up, it saluted him, and Davy lit his pipe by the blaze and went on home, with a piece of sunrise in his pocket (Rourke, *Davy*, p. 243). He had become a veritable god, counterpart or brother to the sun, a shaggy and humorous bringer of light.

TALL TALES

Once Crockett had begun to emerge as a comic legend, the character of Davy was readily adapted to whole stories already in circulation. Writers who knew the stories rewrote them for publication, substituting Davy for the story's original hero, who was often anonymous. Some of these stories were folk tales, so this process is one of the few by which genuine folk elements may have been brought into the legend, and the existence of such stories in the folksy-looking almanacs further accounts for the common misconception that Crockett is a true folk hero. Many of the almanac stories came from the earlier books, and a few of those may have been true folk tales, but most of them were invented by the authors or plagiarized from other literary sources. When assimilation of folk material into the Crockett books or almanacs did occur, the assimilation itself was a deliberate part of an act of writing for publication, so the process cannot be thought of as part of the folklore. That the assimilation was deliberate is further illustrated by the fact that the first and second of the three stories to be discussed here by way of example have been collected by folklorists in variations that do not include Davy Crockett. The first appears to have been adapted to his own image by Crockett himself. The third story, which features a talking raccoon, circulated in print before Crockett became its hero; it represents a rare and fortunate discovery because its history is clear, limited, and well documented.

"A Useful Coonskin" is a story which is popularly known in a version featuring Davy Crockett, and there is a very good chance that Crockett himself initiated the yarn's curious history by telling it to Mathew St. Clair Clarke. The story appears first in an abbreviated form in Clarke's *Sketches*, and next in the first chapter of *Crockett's Exploits*, one of the two chapters based upon the Crockett letters given to Richard Penn Smith by Edward

Carey. The trail of fabrications is amazing: Crockett told it to Clarke, and Clarke published it in an inferior form; Crockett was either reminded of it by reading *Sketches* or decided to bypass Clarke's version by working up a better one; Crockett put it in the letter to Carey, perhaps intending to use it in a future book; Smith copied it straight into *Crockett's Exploits*; the almanac makers and innumerable newspaper editors pirated it from *Crockett's Exploits* and spread it far and wide across the country and across the years.

The yarn nicely characterizes Crockett's style of political campaigning and his ability to pull off a good practical joke. It is another of those stories that deserves to be true even if it isn't; however, it is one of the variations of an enormous and universal set of stories, so it is quite unlikely to be literally true. The story set involves con artists and their games, a generic type that must be as old as storytelling itself. In *America in Legend* (1973), Richard M. Dorson writes that the basic pattern of this particular species of the type takes one of two forms: the con artist sells his mark the same item over and over, or the con artist repeatedly buys something from his mark with the same item of trade, swiping it back each time and using it again. In one variant, a lad in Georgia sells a pair of edible terrapins ten times to the same man. In another, from the Ozarks, two brothers buy supplies at a trading post and pay for them with a coonskin. Later, they buy whiskey the same way. This goes on until the merchant discovers a loose board and a piece of wire at the back of his storeroom and figures out how the brothers were able to retrieve the coonskin for each new trade (Dorson, *America*, pp. 67-75).

Crockett, as the fabricator of his version of the story, may simply have made himself the central character of a folk tale like one of these. On the other hand, maybe he invented it out of whole cloth. Or perhaps it really happened, just as he told it. The *Crockett's Exploits* version is written in the always useful first person, a narrative technique which lets the yarn spinner tell on others by pretending to tell on himself. The story is about how Crockett bought drinks for his constituents during a campaign even though he was flat broke. His mark was old Job Snelling, a crusty Yankee who had originally come out West trading and then had set up a crossroads tavern. Crockett and his opponent were debating at this crossroads, and it came time for Crockett to treat. Job deserved to be tricked, because he had refused to extend credit, a time-honored custom in the West. So Crockett went out and killed a raccoon and came back to trade its pelt for rum all around. Everyone cried "Huzza for Crockett," so naturally Davy wanted to keep them in their good humor. Glancing down, he saw that Job had stuck the coonskin between the logs that held up the bar, and one little end of it was sticking out. He gave it a jerk and slapped it up on the counter for another bottle. Job was fooled, and the trick was good for ten repetitions.

Two results of Crockett's trick reveal quite a bit about frontier sensibilities, even if the story is a hokey fiction. First, the voters admire

Crockett for his deftness and for outwitting the trader who has so often conned them, and they promise him their votes because they believe his shrewdness will work in their favor when he is in Congress. Second, Job forgives the debt Crockett has thus piled up, claiming that it is good for him to be taken once in a while because it keeps him on his toes. In an additional comic twist, however, we learn that Job can afford this show of magnanimity because he did not really lose anything: he charged it all up to Crockett's political opponent.

Job Snelling is missing from the version in Clarke's *Sketches*, and we learn there that using coonskins as currency was not unusual in the West. Clarke says that he does not believe the story to be true, a claim which absolutely proves that he did not know how to handle his best material. We can be sure that Crockett told the story as if it were true, for mock veracity is the first law of yarn spinning, and a declaration of his story's truth is often the yarn spinner's opening line. The liquor in the *Sketches* version is whiskey, not rum, and the nameless merchant tosses the skin up into his loft rather than placing it within reach under the bar. Crockett uses his ramrod to reach up through a space between the logs, twists it in the raccoon's fur, and pulls it back down. (For the texts of both the long version of Crockett's coonskin story from *Crockett's Exploits* and the short version from Clarke's *Sketches*, see chapter 3 of this volume.) The detail of the ramrod makes it entirely possible that Dorson's Ozark folk tale, with its device of the wire, actually has its origin in *Sketches*. On the other hand, there is no reason to believe that Crockett made the story up when he told it to Clarke, rather than pulling it from his memory's stock of traditional backwoods yarns, in which case it may very well be a genuine folk tale.

Dorson also believes that the story "Colonel Crockett and the Bear and the Swallows," which appeared in *The Crockett Almanac* for 1840, is derived from an international folk tale commonly called "How the Man Came Out of the Tree Stump." He reports that field collectors have found the folk variant, without Crockett, in Kentucky, Illinois, Indiana, Missouri, New Mexico, New York, North Carolina, and Wisconsin. Dorson also writes that the Crockett version is "no hack writer's invention" (Dorson, *America*, p. 79). He may be correct in judging the basic story a true folk tale, but the one in which Crockett replaces the anonymous narrator is very much an almanac maker's invention, which may or may not be a hack writer's invention, depending on how the reader chooses to evaluate such things.

Like "A Useful Coonskin," this is a first-person yarn, told by Davy as a story about himself. The narrator pretends to know for certain that the folk beliefs about swallows are all untrue. He says he does not believe that if a farmer kills a swallow, his cow will give bloody milk, nor can he accept the notion that swallows fly south in the fall and come back when the white oak leaves are as big as a mouse's ear. He is telling this yarn to discount the folklore that says swallows spend the winter at the bottom of some pond. He found a flock of them, he says, emerging from an old hollow sycamore tree one spring. His notion is that they must have been in there all winter. Climbing the tree to see what the birds have been doing inside it, he leans too far over the edge of the upper rim and falls in.

Now the narrator turns the satire of the story in upon himself, in the classical pattern of self-derision. Having slipped down inside the swallows' roost, he says, he found himself to be "a little the nastiest critter ever you saw, on account of the swallows' dung." To make matters worse, the sky above him suddenly darkened, and the next thing he knew, a bear had backed down the stump and was about to crush him. Thinking fast, he grabbed the bear's tail in his teeth, drew his hunting knife, and poked the bear in "his posterities" with it. Naturally, the bear climbed smartly back up the stump, dragging the hunter out. Davy concludes by saying that he could have killed the bear, for "there are a wicked sight of virtue in bear's grease," but in all fairness, he had to let the creature go, "for it would be unmanly to be unthankful for the service he done me" (Dorson, *America*, pp. 77-79). The language in which this tale is told is close enough to Crockett's own to sound authentic; whoever adapted it originally did a good job, and perhaps we can judge it a journeyman's invention rather than a hack writer's invention.

A story about a raccoon which surrenders to Crockett, because he knows the hunter's reputation as a crack shot, began circulating in the early 1830s in a version that starred an entirely different character. (Both stories will be found in chapter 3.) In this case, the hero displaced by Davy Crockett was a real person, and the medium in which the story appeared was definitely print. The character was Martin Scott, a young army officer who had gained considerable fame along the Old Southwest frontier for his marksmanship. As a lieutenant, he was stationed at Fort Smith on the Arkansas River, as was another officer, Lieutenant Van Swearengen, whose reputation for being a terrible shot was as large as Scott's reputation for being a good shot. By the time the story circulated among the newspapers, Scott was a captain, so the story became known as "Captain Scott's Coon Story."

In the *Spirit of the Times* for October 13, 1832, the story is told by a correspondent in very formal, correct language that seems to take for granted the ability of animals to talk. A dog trees a raccoon and begins to mock him, declaring to the poor creature that it is futile for him to remain hidden up in the tree, for a hunter will shortly come along and shoot him down. The raccoon politely asks who the hunter might be. The dog replies that it is Lieutenant Van Swearengen. The raccoon knows the man's reputation, so he laughs scornfully and says that Van Swearengen may shoot and be damned. The dog is considerably embarrassed. Sure enough, Van Swearengen comes up, fires several times, and leaves the raccoon howling with glee. Then another hunter can be heard approaching through the brush. Again the raccoon asks who is coming, and this time it is the dog's turn to chortle as he announces Captain Martin Scott. The raccoon cries in despair, for he knows that he is a "gone coon." Folding his paws across his chest, he tumbles out of the tree and surrenders to the dog.

"Captain Scott's Coon Story" is in a letter signed by "An Arkansas Hunter" and was printed by William T. Porter in an earlier issue of the *American Turf Register and Sporting Magazine* as well as the 1832 issue of the *Spirit*. The events in the story are reputed to have happened in 1818 (Yates, pp. 170-73; Blair and Hill, pp. 128-29). A great many writers and correspondents for the *Spirit* referred to it in later years, and the most important literary allusion occurs in Thomas Bangs Thorpe's "The Big Bear of Arkansas." There, Jim Doggett ends his thundering tall tale by saying that he did in fact shoot the great bear, but he had come to think that the final shot had been too easy, and the bear must have been truly unhuntable after all. Jim speculates that the bear "jist guv up, like Captain Scott's coon, to save his wind to grunt with in dying." But then he rejects that speculation, concluding that the bear must have simply died when his time came, which happened to be the moment Jim's shot hit him.

When Crockett was made the hero of this old yarn, his sympathy and fairness were emphasized, while the economy and comic exaggeration of the tall tale itself were preserved. The yarn appeared in a Crockett almanac for 1842 and was posed as having been told by Davy. Having spied the raccoon, Davy takes aim, but the animal raises one paw and asks, "Is your name Crockett?" Davy says yes, and the raccoon volunteers to come on down and surrender. A nice twist occurs at this point, for kindly Davy just naturally has to let the raccoon go, since the little fellow has demonstrated both wisdom and good manners. The raccoon does not stay behind to chat, however; he moves off as quickly as he can, just in case Davy should change his mind. Both forms of this yarn continued to be widely reprinted in the 1840s, but Crockett's is the one that has come down to us.

How did such yarns find their way into the offices of the purveyors of the Crockett almanacs and the writers of the spurious Crockett books? The primary media were undoubtedly newspapers and correspondence. In the 1830s, the American frontier was located in regions not too far away from old cities which had important publishing resources. Weekly papers like the *Spirit of the Times* followed the trade routes across the mountain gaps and down the rivers, out to the West and the Old Southwest, and letters containing local yarns flowed back. Wild country lay within a journey of one to three days from St. Louis, New Orleans, Atlanta, Augusta, and Charleston. Letters from army outposts and small frontier towns could get to the newspaper offices in these cities in the same amount of time. In the taverns, inns, and hotels, reporters and writers swapped stories with travelers fresh in from the backwoods. A journey of four to six days would extend the range of this lively exchange to Cincinnati and Philadelphia. A newspaper containing an editorial call for original stories could go out to frontier towns, and correspondence in response could come back to New York or Boston in less than a month.

Newspaper editors expanded this traffic by marking their favorite stories with "Other papers please copy," and this kept tall tales and hilarious yarns circulating from paper to paper and often back to the papers where they had begun. In the 1840s and 1850s, paperback anthologies of yarns taken from the newspapers became very popular, and the newspapers in turn reprinted yarns from the anthologies. The use of Crockett's letters in *Crockett's Exploits* is an example which shows that occasionally a piece of correspondence went straight into a book before making an appearance in the newspapers. The people who combined old stories with the character of Davy Crockett got all the information they needed from the popular books and periodicals of the day, and in their turn they promoted a whole new popular legend which fed back into the books and periodicals. Somehow, we who live in the age of the media deluge find all this quite familiar.

ACTORS ACTING THE ACTOR

Virtually all of the highly popular early images of Davy Crockett had their beginning in Paulding's play, and the two most sensational avatars of the hero since then have likewise occurred in dramatic productions, one a stage play and the other a series consisting of several television shows and two motion pictures. *Davy Crockett; Or, Be Sure You're Right, Then Go Ahead* was a smash hit production that toured the United States and England from 1872 to 1896, playing continuously to full houses. Walt Disney's famous series appeared in the mid-1950s and generated an enormous but short-lived Davy Crockett fad.

Be Sure You're Right was written by Frank Murdock and staged, revised, and expanded by Frank Mayo, a famous character actor who took the role of Davy. The play is a melodrama, a fine example of an important popular art form in late nineteenth-century America and England. Its fabulous run was terminated only by Mayo's death at the age of fifty-six; he last played the role just two days before he died. Whereas Mayo's portrayal had a life span of two and a half decades, Disney's Davy Crockett, played by Fess Parker, lasted about two and a half years, precipitating in this short time all the predictable by-products of a vigorously promoted and hugely successful media event in the middle of the twentieth century: comic books, story books, a syndicated comic strip, officially authorized food products and toys, and millions of coonskin caps. Both Mayo's Davy and Parker's were noble backwoods types reflecting more of the character of Cooper's Leatherstocking than of the raunchy Davy of the almanacs. Murdock's play, however, has a quirky humor of its own, and Disney's production cultivates, with varying degrees of success, some of the healthier humor to be found in the legend's older artifacts.

The modern editors of Murdock and Mayo's play, Professors Isaac

Goldberg and Hubert Heffner, classify it as one of those melodramas which, it is perfectly safe to say, are "utterly without literary value or literary pretension." The best-known example of this genre of purely entertaining popular theater is Ellen Wood's *East Lynne* (1861), an English play which found a tremendous reception in America. Such melodramas were deliberately designed to appeal to their audience's most spontaneous emotions. The authors and directors employed suspense, pathos, stereotypical characterizations, elaborate pantomimes and tableaux, songs, and lots of unalloyed sentimentality. Another common technique was to veneer the staging with superficial touches of realism: the property list for *Be Sure You're Right* indicates for act 1 "Plenty of dry leaves to cover the stage" and for act 2 "Four buffalo robes" (Goldberg and Heffner, *Lost Plays*, vol. 4, p. ix and p. 116).

Be Sure You're Right is valuable to us now because it shows what the public liked and what artists in the theater took to be a credible representation of backwoods character and idiom. Apparently their audiences, in western towns as well as eastern cities, were far enough removed in time from the old frontier to agree with them. The poverty and hard work inherent in pioneer life are sentimentalized in the play. There is plenty of corny humor, but a lot of it is clearly unintentional. Backwoods speech is presented comically, and a few moments of gentle humor have to do with the antics of children. Davy's lines include a few mild jokes. But the dominant humor of the play is its incongruous mixing of backwoods talk and frontier values with melodrama, sentimentality, and, of all things, a plot involving romantic love.

This Davy Crockett is a bachelor living with his mother, Dame Crockett. The two of them are surrounded by nephews and a scrappy niece named Little Sal. Davy is trying to teach one of the boys how to shoot, but the lad keeps hitting the squirrel an inch from the eye, which makes Davy righteously ashamed of him. Our hero is given to expressions like "Dear me, I forgot them horses!" and "Mighty, but she's pretty!" This last is in reference to the heroine, Eleanor, with whom he is falling in love. In one scene, Davy's heart responds warmly to her reading of "Young Lochinvar" from Walter Scott's *Marmion*. He naturally attributes his feelings to his literary sensibilities: "Well, there's something in this rough breast of mine that leaps at the telling of a yarn like that." Perceiving an analogy between the action of the tale and his own situation, Davy suddenly becomes excited, and this flusters Eleanor. She tries in confusion to apologize: "Heavens, what have I done? Sir—Mr. Crockett—Davy—." He responds, "Oh, don't mind me. I ain't fit to breathe the same air with you. You are scholared and dainty, and what am I, nothing but an ignorant backwoodsman, fit only for the forests and the fields where I'm myself hand in hand with nature and her teachings, knowing no better?" (Goldberg and Heffner, *Lost Plays*, vol. 4, p. 133).

As Blair and Hill have said in *America's Humor*, this second reincarnation of Davy Crockett bears little resemblance to the "crude, bellicose, horse-sensible, tall-talking Davy" of the 1830s, for he is instead "brave, gentle, chivalrous, eloquent and sensitive." They identify as the best example of unintentional humor in the play the scene that closes act 2 and opens act 3. Davy, the heroine, and her useless citified fiancé, who has been laid low by a touch of fever, are trapped in a cabin by marauding wolves. It is the dead of winter. The heads of wolves are seen popping through the windows and under the door. Snow blows through the openings. The heroine despairs, " 'Tis true, nothing can save us!" Davy answers, "Yes, it can!" She asks, "What?" He replies, "The strong arm of a backwoodsman," and bars the door with his arm. This drastic measure is required because Davy has already used the wooden bar as emergency fuel for the fireplace. When act 3 opens, it is nearly dawn, and Davy has been barring the door with his arm all night. He knows that the wolves will scatter when the sun comes up. He is pale and faint. His opening line is "This is getting kind of monotonous, this business is" (Goldberg and Heffner, *Lost Plays*, vol. 4, p. 134). Blair and Hill rightly judge this to be Davy's best line, "truly a thing of beauty" (p. 144).

Another of Davy's speeches, almost as hilarious, captures the spirit of the whole play. The plot involves the usual dilemma of the genre: a cruel manipulator named Crampton is about to ruin the heroine's uncle, who is helplessly and hopelessly in debt to the villain. The heroine, Eleanor, was once, as Dame Crockett tells us, "Little Nell," Davy's childhood sweetheart before she was "took across to forren parts to be eddicated by her father." The uncle is her guardian, her father having long since died. The fiancé, who is fair enough to recognize that he has lost Eleanor's love, but who is also well informed about her family's looming losses, relinquishes his claim upon her. Davy's funny speech occurs when he confronts Crampton to warn him that folks around here don't cotton to villainous strangers:

> Look here, do you know what we do with men like you in these parts, when a man wearing the image of the Almighty Maker shames nature and changes off with the wolf? We of the hills and mountains band ourselves together, and form a court of law where there's mighty little learning, maybe, but where there's a heap of justice, and where a judge sits that renders a sentence—strikes terror to the boldest heart. Do you know his name? It's Lynch—Judge Lynch.
> (*Lost Plays*, vol. 4, p. 147)

When the villain professes not to understand, Davy says, "I mean business, and damned little of that." What this scene draws from the legend is the idea of Crockett as the audacious and irreverent straight shooter who defies the deceptions of convention and punches through to a bold solution. What it recalls from the autobiographical *Narrative* is Crockett's portrait of himself as a frontier magistrate dealing out fair judgments without benefit of "law learning."

As it turns out, it is not Davy the crack shot who finally resolves the melodramatic dilemma. In a nice contradiction of the dictates of the play's own genre, it is the heroine who saves the day. Against her uncle's anguished protest—"Your fortune is sacred!"—Eleanor signs over her inheritance to pay off his debts. Her "forren eddication" has not ruined her, and Eleanor proves true to her backwoods origin. An independent, self-reliant woman, she thus frees herself from all the obligations and encumbrances acquired during her temporary sojourn among civilized—and therefore corrupt—guardians, lawyers, and suitors. She chooses life in the woods over the artificial world money can buy, claims our hero as her own, and prepares to take her rightful place in "the heart and home of Davy Crockett." Mayo's audience—and it must have been an enormous audience if it sustained the play for a quarter of a century—loved every minute of this delightful mush.

Walt Disney understood the mysteries of pleasing a truly massive audience, too. As early as the 1930s, he knew that the infant invention, television, would someday, somehow, become important, and he was always careful to reserve the television rights for his motion pictures. This took considerable foresight, because few people in the film industry—or any industry, for that matter—could conceive of the impending centrality of that fuzzy picture flickering on the tiny round face of a buzzing glass tube. In spite of his prescience, however, he was reluctant to become involved in the television industry itself when it began its enormous commercial expansion in the late 1940s. Of course, by then he had a long history as a major film maker, and he was also busily engaged in fulfilling his life's ambition to build a gigantic, high-technology amusement park. He made an hour-long special for NBC which was broadcast on Christmas day, 1950, and it received good enough ratings that he did another the following Christmas for CBS. But when ABC later approached him with a proposal for a series, he was not especially enthusiastic about the idea. Apparently he thought that television would not prove particularly useful as a form of theater. But he did see its promotional value, and he finally accepted the idea of a television show because he thought it would boost his other enterprises. Disney agreed to do a series, and ABC agreed to invest in the amusement park. The series took its name from the park, and *Disneyland* came to Wednesday night television in 1954. In the fall of 1961, the show moved to NBC and became Sunday night's *Wonderful World of Disney* (Finch, pp. 359-61).

Soon after the show began on ABC, the Disney television team had the idea of doing some sequences about frontier heroes. They considered Daniel Boone and Johnny Appleseed, but finally settled on Crockett, whose story, they proclaimed, was a neglected American folk legend. They were wrong on two counts, of course: the figure of Davy could scarcely be called neglected, and his story as it appears over the years in print is not exactly a folk legend. The pilot footage was shot in Tennessee in just a few weeks, and the studio discovered afterwards that there was only enough film for a little less than three shows. To fill in part of the empty time, the series' composer, George Bruns, and the scriptwriter, Tom Blackburn, hastily put together a seventeen-stanza song titled "The Ballad of Davy Crockett" (Finch, p. 361). In no time at all, every kid on every block in every town in the United States was singing at the top of his or her lungs, "Davy, Davy Crockett, king of the wild frontier." Mention Davy Crockett to any person who was over three years old in 1955, and the automatic response will be a quick rendition of this awesome refrain. Valiant, frazzled Snooky Lanson kept on smiling as he sang it again and again on the *Hit Parade*, a show committed to the live recapitulation of each week's ten best-selling recorded songs. The exaggerated success of the instantaneous hit surprised the songwriters and the Disney corporation as much as it did the nation's long-suffering parents and teachers. (Bruns and Blackburn's was not the first song titled "The Ballad of Davy Crockett." Botkin, pp. 15-16, has the words and music for a very different kind of song bearing the same title and dating back at least to the 1830s. Botkin calls it "a folklorized Negro minstrel song." See chapter 4 for further remarks on Crockett songs.)

The tremendous success of the *Disneyland* presentation led to a motion picture, *Davy Crockett, King of the Wild Frontier*, which was produced basically by condensing and reassembling the original three episodes. The studio had prudently filmed the series in color in spite of the fact it was broadcast in black and white (Finch, pp. 361-62). However, the film was just a little late in coming; by the time it was released, the audience's enthusiasm had already reached its peak, and the film's reception was modest compared to the success of the series (Blair and Hill, p. 146). A sequel, *Davy Crockett and the River Pirates*, went almost unnoticed and has been entirely forgotten.

Disney's Crockett was played by Fess Parker, who really does appear to be six feet four in his stocking feet. Christopher Finch has written that Parker "plays the hero with a rugged dignity which seems completely appropriate" (p. 362). Blair and Hill in *America's Humor*, on the other hand, compare Parker's Davy to Mayo's and find him not only "gentle, chivalrous, sensitive, eloquent, and brave," but "noble to the point of being stuffy" (p. 146). These two judgments bracket the truth without hitting upon an accurate assessment. First of all, the studio's intention with regard to this characterization is quite clear: Fess Parker's Davy was designed as a role model for children. Our collective memory of Crockett returning from his northeastern tour to storm onto the floor of Congress, deliver his impassioned speech about America as a nation for all people, and rip the Indian removal bill in half comes straight from Walt Disney Productions. Second, Fess Parker actually managed a good deal of humor in his portrayal, and the forms of his joking come straight out of the legend. The humor is scrubbed clean for the younger audience. In addition, the Disney team assigned a lot of loose ends from the legend to Crockett's comic sidekick, brilliantly played by Buddy Ebsen, who went on to a long-lived and popular role in television's *The Beverly Hillbillies.*

The spin-offs from the Disney television series were as enormously successful as the show itself. Scores of different Crockett books, old and new, sold as fast as they could be rushed through the presses. A daily syndicated comic strip appeared in over two hundred newspapers. The great symbol of this colossal media event—the coonskin cap—became an indispensable possession to an entire generation of boys and girls. Toy rifles and fringed jackets did well, too. The fad collapsed as quickly as it had begun, and some time late in 1956, as Blair and Hill recall it, all those books were remaindered, and "rifles, coonskin caps, and fringed shirts stuffed garbage cans from sea to shining sea" (p. 146). The three original Crockett shows have been periodically resurrected, however, and no American child has had to reach the age of ten without having an opportunity to see them. They were shown during the composition of this volume in January 1981, on Home Box Office, a subscription service of cable television.

The Home Box Office showing was divided into three segments of forty-five to forty-seven minutes each, which verifies that they were prints of the original footage shot for *Disneyland*, as it is described by Finch in *The Art of Walt Disney* (1973). The installments are entitled "Davy Crockett: Indian Fighter," "Davy Crockett Goes to Congress," and "Davy Crockett at the Alamo." The first stanza of the ballad is used to open each show, while the later stanzas are spread out across the episodes to advance the action and bridge long periods of lapsed time. The realism of the scenes varies in quality. The Indian fights are crude and suffer from all the clichés that were once imposed upon Indians in Hollywood movies. But both the acting and the Washington sets in "Davy Crockett Goes to Congress" are quite well done; this installment is the most professional-looking of the three. Much of the acting in the Alamo segment is wooden, but the sets are convincing and the action well laid out. Davy's death is not shown. He is the last defender, swinging his broken rifle as he is faded out. His sidekick dies valiantly, shot through and through as he struggles to fire a cannon one last time into a mob of Mexican soldiers inside the compound. His last words are, "Give 'em what fer, Davy."

Great care is taken in the series not to make Andrew Jackson look bad. The writers were careful not to take Crockett's autobiographical complaints too seriously. Jackson is played rather well, with plenty of gruff humor, by Basil Ruysdael, who manages to look just like the Old Hickory of schoolbooks. When the story first opens, Crockett is married to Polly, played by Helene Stanley, and he does not remarry after her death (which is reported to Davy in a letter). Elizabeth Patton is left out of the plot. The third segment, the one otherwise afflicted by unremarkable performances, includes a dazzling portrayal which seems to have been overlooked by later commentators trying to recall the series: Hans Conried as Thimblerig, a character invented by Richard Penn Smith for *Crockett's Exploits.* The Bee Hunter in *Crockett's Exploits* is replaced in the Disney series by a speechless Indian, a Comanche reject who could do nothing right as long as he stayed with the tribe. Dubbed "Busted Luck" by Davy's sidekick, this character is played by Nick Cravet. When Davy talks sign language to Busted Luck, Conried says, in his hilariously nasal and impeccably effete style, "Can the Colonel actually converse with that aborigine?" and "I for one do not trust that perfidious savage." Of course Thimblerig and Busted Luck die together as fast comrades in the final scene.

The name of Davy's sidekick as played by Buddy Ebsen is Georgie Russel. In the factual history of Crockett's life, there is a George Russell (with two *l*'s), who accompanies David on a scouting mission in 1813, during the Indian skirmishes. In the *Narrative,* Crockett tells how Major Gibson chose him for the assignment and allowed him to pick his own companion. When Crockett presented young Russell, Gibson said that "he thought he hadn't beard enough . . . he wanted men, and not boys." Crockett answered that if courage were "measured by the beard . . . a goat would have the preference over a man" (*N*, p. 75; see excerpt in chapter 3 of this volume). James Shackford adds that George was the son of Major William Russell, an early settler in Franklin County (*DC*, p. 20). This is all we have on the original George Russell's relationship to Crockett. The very absence of historical detail gave the Disney team license to expand upon this character. The Buddy Ebsen role carries a good deal of the story's humor, and it is his Georgie Russel who has supposedly written the ballad used to pin the show together. Usually the ballad's stanzas are sung by unidentified male voices over the action, but occasionally Ebsen croaks out a few lines while accompanying himself on a mandolin. Most significantly, we learn in the third segment that Georgie has been publishing pamphlets containing tall tales about Davy. It is he who is given the entire credit—or blame—for being the original promoter of the Crockett legend.

Fess Parker's share of the humor is expressed primarily in his use of warped expressions reflecting the spirit of the legendary Davy's talk. He introduces himself to Congress with a version of the famous "I'm half horse, half alligator" speech that is true to the legend. The monologue is related to the formula as it is presented in Clarke's *Sketches* and to several almanac versions which were abundantly reprinted as "Colonel Crockett's Speech to Congress" (see, for example, Botkin, pp. 27-28, and chapter 3 of this volume). Its humor is amplified by the reactions of the puzzled and bemused congressmen. Parker says "not perzackly" for "not exactly" and "flutterated" for "flustered." The legendary story about how Davy grinned the skins off raccoons and the bark off a tree is here expanded into a running joke. He fails in an experimental attempt to grin a bear to death, so has to kill it with his knife; he tries to grin down an Indian, and is surprised when the Indian just gets angrier. Throughout the series, he grins his grin to register amusement, irony, and the awareness of being in a tight spot.

Three exchanges between Parker and a second character will serve to exemplify the kind of joking that the Disney team thought appropriate to their version of Davy Crockett. Early in "Indian Fighter," Jackson commands the militia to initiate a battle by attacking a Creek Indian camp while the regular troops fire from the perimeter. Davy says to the general, "Sure hope your reg'lars don't mistake us for Injuns," and Jackson replies, "If they do, I'll see they apologize." Scouting cautiously around a basin in West Florida, Davy flushes a nest of little reptiles. "Baby gators," he says, "kinda cute, ain't they?" Georgie answers, "Not when they's growed up, they ain't. . . . I'm more scared of snakes and gators than I am of Injuns." Informed by Russel that he has to register his land claim on the Obion River, Davy says, "Where at do we gotta go to do this?" and "This country's gettin' almighty civilized." When he signs his claim, Davy bites his tongue and screws up his face in a painful moment of fierce concentration. These examples are at least as good as most of the jokes in the traditional artifacts of the legend, and they are far more palatable than the humor of the later almanacs. The inescapable conclusion is that the Disney series is a major contribution to the legend, even if it is perceived by some scholars as a popularized distraction from the factual history (for instance, see Folmsbee in the introduction to *N*, p. x, fn. 1 and p. xv, fn. 19).

Fess Parker's Davy dominates our contemporary ideas about Crockett, but the film portrayal most faithful to the many dimensions of both the legendary figure and the man is John Wayne's role in the motion picture *The Alamo*. This movie was produced and directed by the Duke himself and released through United Artists. Wayne, a legend in his own right, was supported by remarkable performances from Richard Widmark as James Bowie, Laurence Harvey as William Travis, and Richard Boone as Sam Houston. James Edward Grant is credited for the screenplay. The film made a modest run in 1960 and then was released twenty years later to be televised by ABC on the Fourth of July, 1980.

Wayne's *The Alamo* is far from a great picture; in fact, some of the scenes are downright embarrassing. But Wayne himself is well worth watching while keeping in mind the history of the Crockett legend and the difficulties of retrieving some of the facts of Crockett's life and death. Wayne captures Crockett's sense of humor, his clumsy gallantry and sentimentality, his impatience and his egotism. He displays Crockett's moral flexibility, his harmless streak of trickiness, and his lusty bragging. He shows us Crockett's moderate love of flirting, rowdy horseplay, and whiskey. The

Tennesseans riding with him sometimes wear coonskin caps or beaver top hats, and Wayne occasionally does, too, but he usually wears a wide-brimmed, flat-topped cloth hat.

Acknowledging the various historical problems, the film shows Crockett as being aware that the Texans are planning an independent state, and Wayne delivers an awful speech on the meaning of the word "republic" that probably extrapolates Crockett's sentiments quite well. Travis says he thinks Crockett's talk, dress, and mannerisms are merely a pose. Then he accuses Crockett of coming to Texas just to get into the fight. Wayne answers, "Don't tell my Tennesseans that—they think they came to Texas to hunt and get drunk." He then claims to have come to the Alamo to fight for what he believes to be right. The hero in this film is seldom called Davy except by his closest friends among the Tennesseans; he is most often called Crockett, which matches the common term of reference for him in the newspapers of his own day. Wayne's solution to the problem of Crockett's death was to invent a new scenario for it. Running across the compound with a torch in his hand, he is trapped by Mexican soldiers against the door of the room in the mission where the Texans have stored their powder. The soldiers perforate his body with their lances. He staggers back, throws the torch into the room, and plunges through the doorway himself. We see only the explosion.

The makers of this film were extremely conscious of the public's fresh memory of Fess Parker as Davy, and it is easy to see that they tried to offset it by cultivating all the various known qualities of Crockett's character. They succeeded largely because John Wayne simply took the role over, commanding it by the power of his own personality. His Crockett is much more complex than Disney's. It is probably the most truthful of all the theatrical portrayals of Crockett. John Wayne's legendary image seems, in retrospect, to have fit the legendary image of Crockett quite well. Perhaps this tells us something about both legends.

TRUTHFUL FICTIONS

John Wayne's portrayal of Crockett is truthful because the actor drew upon the facts of Crockett's life and some carefully evaluated details from the legend in order to imagine a faithful characterization. Facts alone would never suffice; actors speak of "getting into" a role or "becoming" a character. Since both John Wayne and David Crockett were distinctive examples of the hard-hitting, straight-shooting, plain-talking American type, the portrayal probably came to the actor quite naturally. Some creative process would likewise inform a truthful Crockett book, which would not necessarily be a biography. Good factual accounts, like James Shackford's scholarly biography or Dan Kilgore's carefully researched monograph about Crockett's death, give us the necessary historical details about the man's life and a sense of the historical context in which those details are embedded. A history is a narrative, or story, and it is fictional to the extent that the historian cannot tell the whole truth because he is restricted to the surviving facts and a highly selective presentation of the historical context. He also invents a way to tell what he has found. But the historical biographies of Crockett display little of the process of getting into the character and his times. A good biographer can get into the character of his subject when he has plenty of definitive and revealing material, but the words "plenty" and "definitive" do not describe the factual Crockett record.

Perhaps a good fiction, presented consciously and conscientiously as a fiction, could tell a faithful story of Crockett. It would draw upon the facts and those features of the legend which are true to Crockett's character and to his imagination. Disney's production has some of the qualities of such a fiction; in a different and more interesting way, John Wayne's role as Crockett is quite a good one. Is there a truthful Crockett fiction in the form of an imaginative prose narrative? There are several good narratives which were not consciously intended to be perceived as fictions. There are fictions consciously derived from the legend, but most of them are mediocre and a few are atrocious. But there is one very deliberately crafted novel, to be discussed last, that deserves the title of the most truthful Crockett book.

All of the artifacts of the Crockett legend, as opposed to the few serious biographical items, are deliberate fictions of one kind or another. Some draw upon the facts, but in every case the author's or producer's design dominates the story. Occasionally this design is better called an ulterior motive. The fictions display various degrees of imagination. The almanac yarns, for instance, draw upon other sources, some fabricated in print shops and a few genuinely representative of folklore, but there is virtually no imaginative transformation of the borrowed material in the almanacs. The *invention* of the Crockett almanacs, especially the Nashville series hoax, is an imaginative act, however. The early almanac illustrations also represent an artistic transformation which is important to the legend. The quality of imagination that went into Paulding's *The Lion of the West* is high; the inventiveness of Clarke's *Sketches* and Smith's *Crockett's Exploits* is ordinary, and typical of such books at the time.

Crockett's own *Narrative* is an imaginative book; furthermore, the imagination of its author is one of its main subjects. The *Narrative* is more faithful to the facts of Crockett's life than any other contemporary document, but it is also a transforming act, for a good deal of invention went into the process of selecting what would go in and what would be left out, and into the fabrication of an appropriate style. Crockett, with Thomas Chilton's judicious assistance, designed a prose narrative that would evoke a sense of his backwoods attitudes and speech, and this is what makes the book a good portrayal of Crockett's imagination. It recapitulates his style of talking, his storytelling, his beliefs, and his way of promoting his own image.

Most of the books about Crockett that have appeared since his own time draw heavily upon the *Narrative*, Clarke's *Sketches*, Smith's *Crockett's Exploits*, and the Whig-sponsored *Colonel Crockett's Tour*. In fact, one of the most common books to be found in libraries combines the *Narrative, Colonel Crockett's Tour*, and *Crockett's Exploits* into a single running narrative that is posed as Crockett's autobiography or as an edited and annotated edition of his writings. One undated edition is titled *Life of David Crockett, the Original Humorist and Irrepressible Backwoodsman*, and was published by Lovell, Coryell & Company, New York. Another has the same title plus the word *The* at the beginning and the words *An Autobiography* at the end. It was published by A. L. Burt, New York, and dated 1902. A popular edition published in 1934 by Charles Scribner's Sons contains paraphrases of the *Narrative* and *Crockett's Exploits*, and is titled *The Adventures of Davy Crockett, Told Mostly by Himself*.

Readers interested in either David the man or Davy the legend cannot afford to be casual in their approach to these variously combined editions. Richard M. Dorson cites an undated printing of the A. L. Burt edition and calls it Crockett's *Autobiography*, thus confusing it with the *Narrative*; the item Dorson quotes is from the *Crockett's Exploits* section of the book (Dorson, *America*, p. 70, and p. 313, fns. 15 and 16). This "life" is a good example of the many random reprintings and recombinations of the Crockett books that have been published over the years. Sometimes a recombined reprinting contains the name of "Alex J. Dumas" as the writer of a preface which quotes a letter written from Texas by one Charles T. Beale. As Shackford has demonstrated, Alex J. Dumas never existed, and the letter from Beale, who was a traveler in Texas, is a pure invention (*DC*, pp. 273-78). These devices are residual from the first edition of Smith's *Crockett's Exploits* and are parts in the machinery of his fabrication of the Crockett diary.

One such recombined edition has exceptional literary value because of the identity of its editor. This is *The Autobiography of David Crockett*, assembled and introduced by Hamlin Garland and published by Scribner's in 1923. It includes the *Narrative, Colonel Crockett's Tour*, and *Crockett's Exploits*. Garland knew the bibliographical histories of the books quite well. He dismisses Clarke's *Sketches* as "an unauthorized work which its subject resented," but includes *Crockett's Exploits* for reasons that tell us something about Garland's own literary perceptions:

> It has seemed worth while to reprint this pseudo-Crockett, except for the preface and the final chapter, not only because it is itself interesting but because the existence of such a book shows that there was current at the time a popular legend and literature of the frontier

which made it possible for catchpenny hacks to manufacture a reasonably characteristic, reasonably convincing "autobiography" of a dead hero while his death was still in the news.

(p. 11)

Garland knew that Alex Dumas and the Beale letter were fakes, and this is why he omitted the preface and conclusion from his edition.

In some very interesting introductory comments, Garland expresses his authorial consciousness of the importance of the authentic representation of speech in an American prose narrative. He writes that Crockett's use of expressions like "I know'd" and "in and about" matched those he heard used "by the old men of Wisconsin in 1868." Garland did not know the exact history of the composition of Crockett's *Narrative*, but he was sure that it accurately portrayed the hero's "naive boasting and homely humor" (p. 4). He recognized that what made Crockett a successful campaigner was "the strength of his arm, the keen glance of his eye, and his abounding humor" (p. 8). One of Garland's assessments of Crockett and the *Narrative* is absolutely brilliant: "That he wrote it as it stands is doubtful, but that he talked it is unquestionable" (p. 4).

Garland's text has a peculiar place in the long history of Crockett artifacts. It stands as a link between Murdock and Mayo's *Be Sure You're Right* and Disney's Davy Crockett series. As Stanley Folmsbee points out, Garland's edition was extremely popular, and its appearance guaranteed the perpetuation of the spurious stories begun by Richard Penn Smith and the purveyors of *Colonel Crockett's Tour*. These stories are important to the second and third segments of Disney's series. As a historian, Folmsbee objects vigorously to Garland's inclusion of spurious material (*N*, p. v). The student of popular culture may take a more positive view. The link backwards in time to Murdock and Mayo's play is in Garland's introduction, which begins with this marvelous reminiscence:

> Many years ago, while a student in Boston, I saw Frank Mayo play the title role in Murdock's comedy "Davy Crockett," and though I greatly enjoyed it, I can recall but two scenes: one in which the young hero sadly confesses to his sweetheart his inability to read and write, and the other an exciting moment wherein a band of wolves gnaw their way through the wall of the settler's cabin with such incredible ease that I wondered what kind of logs had been used in its construction. The first scene was altogether charming. The soft-voiced, bashful, handsome young hunter in his fringed buckskin jacket and coonskin cap quite won the hearts of the audience, and we were all grateful when the girl of his adoration offered to teach him his alphabet. The playwright had made of Davy a young Lochinvar of the Canebrake, endowing him with all the romantic virtues. He was

chivalrous, generous, and a poet, the prototype of the long line of heroes, hunters, cowboys, and miners who from that day to this have filled a large place in our literature.

Garland goes on to say that his reading in preparation for assembling the edition destroyed his romantic image of Crockett as "Lochinvar of the Canebrake." He concludes that "Somewhere between the coarse, bragging, stump-speaking politician and the Davy Crockett of Murdock's play lies the real backwoodsman, whose fame is united with that of Daniel Boone as our typical pioneer" (p. 3).

Occasionally, a person has paraphrased the Crockett books to make a text that is supposed to look like a historical biography. This is what John S. C. Abbott did for his notorious *David Crockett*, published in 1874. Abbott acknowledges his sources in curious ways. Once, he states that he intends to present Crockett's "peculiar character exactly as it was" and defensively tells us, "I have therefore been constrained to insert some things which I would gladly have omitted" (p. iv). In other words, some of the things he plagiarizes embarrass him. This is why he is careful when he quotes them: "I omit the profanity, which was ever sprinkled through all his utterances" (p. 33). Abbott's spurious technique and the inaccuracies of what little material he added are pointed up by Emerson Hough in his important book *The Way to the West* (1903), which includes biographical information on Crockett, Daniel Boone, and Kit Carson. Hough was amused by Abbott's assertion that Crockett had killed a great many ferocious grizzly bears (Hough, p. 163). Crockett's quarry was, of course, the smaller black bear.

Shackford is quite right in assigning most earlier biographical work to the history of Crockett's legend rather than to the history of his life and times. The classic discussion of the difficulties posed by the scarcity of historical facts and the abundance of fabricated material is Walter Blair's essay "Six Davy Crocketts." Blair identifies a historical David Crockett, but he thinks this is the least interesting one. The fabrications include the image promoted by the newspapers, the almanac Davy, the political caricature which made him a fool, and the political caricature which made him a wise rustic. The fully developed legendary figure is the one Blair likes best, for this is a demigod, the type of the western hero, and it evolved as a collective invention—the people's Davy Crockett.

But this proliferation of fictional Crocketts is no cause for cynicism. If the historian is baffled or frustrated, the student of popular culture has struck it rich. Several of the best fictional histories about Crockett can be perceived as important contributions to the legend, and this is true even of the scholarly books written about Crockett before the publication of Shackford's biography. In other words, serious investigators who set out to track Crockett have often ended up by expanding the legend. In turn, we ought to remind ourselves that it is the expanding legend that gives the careful historian like James Shackford or

Dan Kilgore the motivation for pursuing the life of Crockett, for the facts alone would attract few scholars.

In the intriguing category of fictions which have taken the place of biographies, we can include Charles Fletcher Allen's *David Crockett, Scout* (1911), Irwin Shapiro's *Yankee Thunder* (1944), Stewart Holbrook's *Davy Crockett* (1955), Edwin Justus Mayer's *Sunrise In My Pocket* (1941), and Edward S. Ellis's *Life of David Crockett* (1884), a biography which has special significance largely because of the rest of its author's career. Allen pretends to follow Crockett's actual history "with close attention to dates, and without recognition of the impossible legends of many writers" (p. vii). Naturally, his book relies heavily upon Crockett's own tales and Smith's *Crockett's Exploits*. An interesting feature of the volume is the inclusion of Chapman's bust portrait of Crockett in a Byron collar.

Shapiro wrote his book primarily for children, and his introduction reveals his consciousness of the historical problem. After some searching, he decided it was the legend that fascinated him and would fascinate others, so he discarded the factual David and wrote a book about Davy. Holbrook's is likewise a children's book. It is based partly on the facts and partly on the legends; its index and explanatory material make it look like a biography. Curiously, its dust jacket shows Crockett fighting Indians while a cabin in the background burns, and this Davy looks mighty like Fess Parker (the publication date places the book squarely in the era of the Disney fad). Mayer's book is written as a drama, but is to be classified as a fiction because—as Carl Van Doren writes in its preface—it "is no tract for the theatre or the screen." The reason, Van Doren says, is that "It is a full-bodied comic story, both picaresque and poetic, in the grand tradition of Cervantes and Fielding and Mark Twain" (p. 4). These are nice compliments, but Mayer's book is actually "no tract for the theatre" because it is simply awful. It is inflated, sentimental, and ridiculously inaccurate with regard to either history or legend. At the final curtain, stage directions call for a voice speaking in a tone "Great, strong and beautiful" and saying, "Thermopylae had its messenger of defeat; the Alamo had none." Mayer probably got this line from Ellis's book, where we are told that it was a motto chiseled upon the cenotaph in the Texas state capitol, which was destroyed by fire in 1881.

Ellis's book is part of a larger context involving the author's career as a contributor to a popular American genre that includes quite a few derivations from the Crockett legend. Edward Ellis was a prolific writer of those much-admired and much-maligned—but mostly much-read—fictions called dime novels. Dime novels only sometimes cost a dime: we call them by this name now, but some of them were nickel novels, some were half-dime novels, some cost fifteen or twenty cents, and quite a few of them were serialized in popular weekly periodicals which specialized in the sensational romance. All were extremely popular and extremely vulnerable to the ravages of time, having been bound in cheap paper covers. Ellis's *Life* is fairly reputable, as these things go, and it sticks closely to the Crockett books, especially the *Narrative, Colonel Crockett's Tour,* and *Crockett's Exploits.* He was widely known for his books on Daniel Boone, and he was responsible for at least two and possibly three sensational Crockett thrillers from Beadle and Adams. *The Texan Trailer, or, Davy Crockett's Last Bear Hunt* (1871) appeared over the name of "Charles E. Lasalle," a known pseudonym of Ellis's. The setting is East Texas just before the battle at the Alamo. Lasalle's name is on Beadle's Boy's Library number 139, *Col. Crockett, the Bear King*; and another book, attributed to "Harry Hazard" and titled *The Bear Hunter, or, Davy Crockett as a Spy,* may also be Ellis's work. (It is also possible that "Harry Hazard" was another dime novel writer, Joseph Edward Badger.)

The Crockett dime novels derive from the Crockett books of the 1830s and tend to emphasize the more adventurous aspects of his story, especially bear hunting, Indian fighting, and the Alamo. Legend and wholesale fabrication far outweigh the facts; surely no one would expect mundane historical actualities among the pleasures and passions of the dime novel. Copies are hard to come by today, but at least a dozen separate titles from several publishers have been identified. Some of these, however, are reprints under different titles and over different pseudonyms (Rourke, *Davy*, p. 259; Leithead; Johannsen).

Constance Rourke pursued the legendary Crockett deliberately, incorporating fact and legend into a truthful fiction which remains one of the basic Crockett texts. Rourke read scores of books and documents from the period and worked into her narrative a wealth of yarns, details, and references which give it the tone of the humor of the Old Southwest. It is, essentially, an authentic picture of one of the matrixes from which the Crockett legend rose. Like Rourke, Walter Blair sought to tell the story as if it were half history, half legend, and he chose to do so in a prose style which suggests the rhythms of backwoods talk. The book, published in 1955 (the year of Disney's Davy), is titled *Davy Crockett—Frontier Hero: The Truth as He Told It—The Legend as Friends Built It.* Like Shapiro's, it is written for the younger reader. Other narratives designed specifically for children and likewise drawing on the legend are *Chanticleer of the Wilderness Road* by Meridel Le Sueur (1951), and *The Story of Davy Crockett*, by Enid Meadowcroft (1952). Elizabeth Moseley's *Davy Crockett: Hero of the Wild Frontier* (1967) is a large-print book for beginning readers.

As a character, Crockett has walked through countless fictions published since the early 1830s. As previously mentioned, Montgomery Damon in Caruthers's *The Kentuckian in New York* (1834) and Earthquake in French's *Elkswatawa* (1836) are modeled after Crockett. Robert Montgomery Bird drew upon Crockett's image for Roaring Ralph Stackpole in *Nick of the Woods: Or, The Jibbenainosay* (1837). Elements of Crockett's fighting challenge are used by William Gilmore Simms to represent

Crockett at the Alamo in the play *Michael Bonham: Or, The Fall of Bexar* (1852), and by John Pendleton Kennedy as part of his characterization of the blacksmith hero of *Horse-Shoe Robinson*, published in 1835 and dramatized for the stage in 1856 (Arpad, "Fight Story," p. 164; Arpad, *A Narrative*, p. 34).

Very interesting allusions sometimes appear as glimpses of the legendary figure in major American books, like Twain's use of tall bragging in the story of Little Davy's triumph on board the keelboat. Herman Melville, in *The Confidence-Man* (1856), gives us a lanky "Missourian," also called a "Hoosier," who wears a bearskin jacket and a coonskin cap, talks tall, argues for skepticism, and emphasizes his points by pointing his rifle. Once, he turns defiantly towards the Confidence-Man and clicks his rifle-lock "with an air which would have seemed half cynic, half wild-cat, were it not for the grotesque excess of the expression, which made its sincerity appear more or less dubious" (p. 92).

The most truthful book of fiction about Crockett is Dee Brown's *Wave High the Banner*, published in 1942 and subsequently ignored by Crockett folklorists and historians. The subtitle states the author's conscientious intentions: *A Novel Based on the Life of Davy Crockett*. As a writer of southern and western history, Brown won wide recognition for a later book, whose underlying methodology was an imaginative act of elegant simplicity. This was *Bury My Heart at Wounded Knee* (1970), a narrative based on the records of treaty negotiations and other documents in which the words of the Indians were written down, especially during the last half of the nineteenth century. The elegantly simple idea was to let the Indians speak for themselves. Likewise, the Crockett novel grew from an elegantly simple idea: Brown supplemented and built upon the facts in order to create an imaginatively true speculation about what could have happened along the route of Crockett's life.

Brown's story depends first upon Crockett's *Narrative* and second, with more caution, upon *Colonel Crockett's Tour* and *Crockett's Exploits*. These books establish the route of the plot. Around the facts and suggestive legendary details are gathered the thoroughly reconstructed dimensions of Crockett's world. An energetic historian as well as a yarn spinner, Brown drew upon a great many documents which provided details about the settings: the various locations in Tennessee from 1802 to 1835, Arkansas and Texas in 1835 and 1836, and Washington City at the time Jackson became president. The book opens with young David's hike across the bleak countryside, the return to his father's tavern in March 1802, and his surprised family's excitement. In the tavern's murky interior, hard-drinking wagoners slump across the tables in a stupor, waiting to be fed. Supper is boiled salt meat, potatoes, squirrel broth, corn bread. The scenes in Washington are especially rich. Here is Crockett in the dining room of the Indian Queen, talking to Anne Royal, a writer of American travel books and a noted gadfly journalist. Crockett, Senator John Henry Eaton, and Eaton's wife, the spunky, colorful, and often slandered Peggy O'Neale Timberlake, try to protect the new president, Andrew Jackson, from the boisterous and sometimes dangerous backwoodsmen who have come to celebrate the inaugural. In the Capitol Bar, Crockett confronts Henry Clay, "hated Judas of the West," while Congressman James K. Polk slips out the nearest door.

If all this sounds like the predictable stuff of historical novels, it is successfully tempered and balanced by its historical accuracy and the author's ability to evoke a strong sense of time and place. A West Tennessee doctor slaps at a mosquito while puzzling over Polly's fever; the streets of Washington are gulleys of mud dotted with happy pigs. After his last defeat, Crockett speaks to his constituents in the town hall of Jackson, "the town that had been a wilderness of canes a few years ago." Stinking oil lamps light the room. " 'I'm done with politics,' " Crockett says to his neighbors, " 'You folks can all go to hell, I'm going to Texas' " (pp. 285-86). The old joke is from one of Crockett's chapters in *Crockett's Exploits*; the details of place and time are from history; the authentic evocation of the hero's character comes from the storyteller's understanding.

THE CONTINUING INVENTION OF CROCKETT

Since Davy Crockett has long been a subject for folklorists, it may be surprising that little of the Crockett legend can actually be labeled folklore. There are some problems inherent to the study of folklore that are magnified in the study of Crockett lore. Any yarn appearing in an old text may or may not have come from an oral source. A Crockett yarn appearing in a newspaper or almanac of the 1830s, 1840s, or 1850s almost certainly did not come directly from an oral source, but could have come indirectly from an oral source. Today, if a folklorist were by chance to collect a Crockett yarn on some front porch in the Ozarks or the Smokies, he would be wise to inquire about what his contact has been watching on television. The woods are full of aerials.

In Crockett's time, yarns traveled east through the Appalachian passes to the major publication centers, or down the rivers to the big cities on the Ohio and the Mississippi, or through the woods and swamps to the old towns of the Southeast. The "aerials" were the almanacs, cheap paperback anthologies, and periodicals, especially the newspapers. These media harvested stories from each other as well as from living yarn spinners. Printed stories followed the trails from the cities back out to the woods and prairies. A few of the yarns may be rightly called folk tales, but they would have traveled a long a curious route before they found their way into print as Crockett stories.

When John Seelye worked out the mechanisms of the history of the Crockett almanacs, the last strong possibility for defining one of the major artifacts of the Crockett legend as pure folklore collapsed. On the other hand, the legend cannot be separated from the hopes, fears, and beliefs of the nation in which it arose and the audience to which it still appeals. This legend is a very strong example of what is meant by the term "popular culture." The stories of which it is constituted are widespread and appear in many different forms. These forms are often dictated by the more useful communication technologies in existence at the time the stories are reassembled and disseminated. A Disney film is a Disney property, and John Wayne's portrayal is controlled by United Artists, but the Crockett legend itself cannot be thought of as having been copyrighted by anybody. Its many stories have many authors, some of whom will remain forever anonymous, and the most important medium in which the Crockett legend exists is the imagination of its audience.

The clue to the life of the legend lies with the people who over the last century and a half have read the papers, pamphlets, and books, bought the tickets, and tuned in the sets. They have demonstrated their sympathy with Crockett's wacky humor, his defiance, his struggle for self-reliance, his irreverence for authority, his dark streak of impulsive and occasionally nihilistic rejection of conformity—that ready willingness to say "You folks can all go to hell, I'm going to Texas." This audience loves nothing more than the sharp crack of a satirical remark which hits its target like a rifle shot. Our reception of his legend is all the more significant because it reveals our continuing awareness that America is an invention, a colossal tall tale that was thought up in Europe, acted out by immigrants, and expanded by westward-moving settlers. The *legend* of Crockett is an inseparable part of the *myth* that is the American West.

This discussion of David Crockett's life and legend began with the contention that the epic yarn titled "Crockett" has its truest life in an indefinite, fluid story which changes and expands across time as it is reused in innumerable ways by an innumerable audience. The various creators of the yarn, from James Kirke Paulding, James Hackett, Mathew St. Clair Clarke, Thomas Chilton, and Richard Penn Smith down to Dee Brown, Walt Disney, and John Wayne, have given it substance and passed it on. But they, too, are participating as members of the audience. This is even true of Crockett himself, who wrote about himself writing about himself, who bowed to his own image in the Washington Theater, and who probably died while living up to his legend. Such a story is never recorded in a final form. It is a collective enterprise, a truly public property.

1. Oil portrait of Congressman David Crockett in hunting garb, by John Gadsby Chapman, painted in 1834. *Iconography Collection, Humanities Research Center, the University of Texas at Austin.*

2. John Gadsby Chapman's romanticized bust portrait of David Crockett, painted in 1834, in preparation for the preceding full-length portrait. *Courtesy, Library of the Daughters of the Republic of Texas at the Alamo, San Antonio, Texas.*

3. A contrasting view of David Crockett. Engraving by Chiles and Lehmann, Lithographers, Philadelphia, 1834, from an oil painting by Samuel Stillman Osgood. *Tennessee State Library.*

4. James Hackett as Nimrod Wildfire in *The Lion of the West*. Engraving made from a portrait by Ambrose Andrews. Captioned: "Come back, stranger! or I'll plug you like a watermillion!" *Courtesy Harvard Theatre Collection.*

5. Woodcut portrait of Davy apparently copied from Andrews's portrait of Hackett. Cover of *Davy Crockett's Almanack*, 1837. Reproduced by permission of The Huntington Library, San Marino, California.

6. Fess Parker as Davy Crockett and Buddy Ebsen as Georgie Russel in *Davy Crockett and the River Pirates*, 1956. © *Walt Disney Productions.*

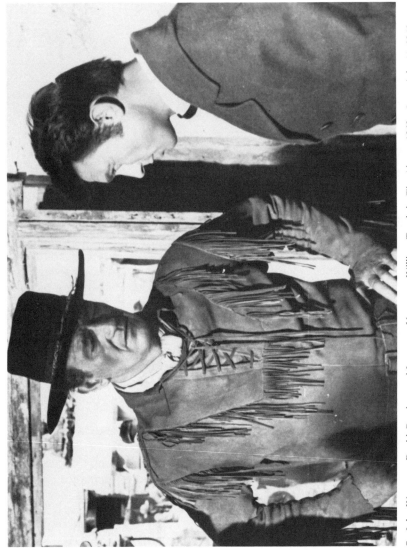

7. John Wayne as David Crockett and Laurence Harvey as William Travis in *The Alamo*, 1960. Copyright © 1960, The Alamo Company. All rights reserved. Released through United Artists Corporation.

3
THE CROCKETT IDIOM

David Crockett's own way of talking and telling stories is displayed best in his autobiography and is strongly suggested in other writers' re-creations of yarns that seem likely to have been his originally. The samples which follow exemplify the printed record of Crockett's talent for contributing to his own legend. They tell us something about his conscious art and give us a picture of the self-conscious humorist at work. Crockett speaks for himself in his letters, of course, but his letters seldom reveal very much about his humor or his command of the backwoods idiom. He usually wrote his correspondence in what he perceived to be a formal style, unless he wanted to include a deliberately prepared story. One of the stories in this selection, the version of "A Useful Coonskin" taken from *Crockett's Exploits*, did come originally from a Crockett letter, by way of Richard Penn Smith's promotional maneuver, as explained in chapter 1.

It may seem regrettable that Crockett's best yarns come to us in contexts that cannot be called pure Crockett. Every good sample has its existence partly because some second party had a hand in it. Among the selections which follow, those from the *Narrative* are to be credited in part to Thomas Chilton, and Mathew St. Clair Clarke can be thought of as co-author of the first version of "A Useful Coonskin," a yarn Clarke attributes to Crockett in *Sketches*. But Crockett's contributions to American humor would not have survived for us to read if the yarns he spun had not somehow found their way into print, and they would not have found their way into print on Crockett's initiative alone. There is nothing in his character to dictate a literary impulse; there is nothing in his background or accomplishments to serve as a motive for recording his own art.

Moreover, Crockett's art is inseparable from its original context, and neither his art nor its context can be closely associated with printed literature. His art was a form of ephemeral theater—oral storytelling for small groups. The stage was a stump in the woods, the floor of a general store, a table in a tavern, or a political podium. One of the most carefully cultivated assumptions of this kind of art is that it exists in defiance of all things authoritative or formal. The backwoods style of yarn spinning was thought of as an antidote to literature, just as its subjects were always posed as being antithetical to established institutions, and just as its heroes always behaved in ways that mocked accepted behaviors. (The humor of stand-up comedians still displays this satirical spirit.) The fact that Crockett had help in creating the printed record of his idiom is thus fortunate rather than regrettable.

Of course, a printed record cannot substitute for an oral performance. If the writers and promoters who recorded the Crockett legend had tried to recapitulate every detail of pronunciation, the tales would have been unreadable. (Many early comic writers in America tried to represent dialect phonetically, and as a result their stories are often extremely difficult to read—Joel Chandler Harris's Uncle Remus books are good examples.) If the writers and promoters had also tried to include thorough descriptions of Crockett's mannerisms and gestures, we would be too bored to pursue the yarns. The following samples show that the creators and re-creators of the legend had various degrees of success in the art of judicious suggestion. The highest achievement is Crockett's, for the *Narrative* abundantly displays grammatical structures and specific idioms which Americans associate with backwoods yarn spinning. An expert in American dialects might find that Crockett's language, as it is represented in print, resists being classified with respect to geographical areas, but the reader who reads these yarns for fun would immediately think of Crockett's style as being exemplary of American tall talk.

The Crockett stories selected for this chapter have been edited for readability. A good many obvious errors in punctuation and spelling (not the kind of misspelling designed to enhance the idiom) afflicted the first editions of Crockett's *Narrative*, Clarke's *Sketches*, and Smith's *Crockett's Exploits*, and many of these were corrected in later editions that contained new errors, so there is no reason to believe a definitive edition of any of these three books exists. In the introduction and chapters 1 and 2 of this volume, quotations from Crockett's *Narrative* are used to establish historical information and arguments, and those quotations follow Shackford and Folmsbee's edition in every typographical detail. Here, however, chapter numbers are given for the selections from the *Narrative*, Clarke's *Sketches*, and *Crockett's Exploits*, so that the reader who wishes to read them in context can consult any available edition.

GEORGE RUSSELL

As we saw in chapter 2, the Disney corporation virtually invented a sidekick for Davy Crockett, so that in 1954 the old legend gained a new and interesting comic dimension. The sidekick is named Georgie Russel, and the role is played by Buddy Ebsen. The following excerpt from the *Narrative* constitutes David Crockett's one reference to the original George Russell. The magnitude of the invention that went into Ebsen's role will be immediately apparent to anyone who has seen Disney's Davy Crockett series.

The passage is important for other reasons. We have here a glimpse of Crockett's humorous defiance of authority and his impudent way of deflating pompous assumptions. Though brief, the passage contains a wealth of idioms, including "I know'd," which every American takes to be a backwoods usage; "I was a little wrathy," which is strongly associated with Crockett's comic image; and "cut out," which may not be Crockett's coinage, but is surely Crockett's in the sense that he made it his and then planted it deeply in our common language.

The locale of the following anecdote is Beaty's Spring, south of Huntsville, and the major whose undemocratic elitism invites Crockett's sarcasm was John H. Gibson (*N*, p. 75).

> While we remained at the spring, a Major Gibson came, and wanted some volunteers to go with him across the Tennessee river and into the Creek nation, to find out the movements of the Indians. He came to my captain, and asked for two of his best woodsmen, and such as were best with a rifle. The captain pointed me out to him, and said he would be security that I would go as far as the major would himself, or any other man. I willingly engaged to go with him, and asked him to let me choose my own mate to go with me, which he said I might do. I chose a young man by the name of George Russell, a son of old Major Russell, of Tennessee. I called him up, but Major Gibson said he thought he hadn't beard enough to please him,—he wanted men, and not boys. I must confess I was a little nettled at this; for I know'd George Russell, and I know'd there was no mistake in him; and I didn't think that courage ought to be measured by the beard, for fear a goat would have the preference over a man. I told the major he was on the wrong scent, that Russell could go as far as he could, and I must have him along. He saw I was a little wrathy, and said I had the best chance of knowing, and agreed that it should be as I wanted it. He told us to be ready early in the morning for a start; and so we were. We took our camp equipage, mounted our horses, and, thirteen in number, including the major, we cut out.
>
> —*A Narrative of the Life of David Crockett* (1834), chapter 5.

THE TALL TALE REVERSED

Tall tales always exaggerate the hero's accomplishments, and one of the most common traditional yarns is a story in which a hunter brags about how his skill produced some incredible harvest of game. In some variations, often classified as one story called "the wonderful hunt," the hunter fires a single shot that kills the deer he aimed at and then, having passed through the animal, hits a rabbit, a flying duck, and a leaping fish. The ball finally punches a hole in a bee tree, and the hunter's bag is further weighed down by several pounds of fresh honey. In backwoods humor, the antithesis of this tall tale is also common, and yarn spinners like to tell of failures which seem just as incredible as the wonderful hunt. To this day, a common story in the Florida Panhandle and South Alabama, the scene of the following excerpt, tells of a party of hunters who set up their fall camp without bothering to bring in meat: they are boastfully confident of their ability to live off the land by eating some small portion of the game they intend to kill. By the third day, they are happy to eat snakes and songbirds.

This episode from Crockett's Indian-fighting experience is perfectly literal, however, and its humor is strictly a function of his style: after the main part of Jackson's army moved from Pensacola to Mobile, the troops who were left behind to chase renegade Indians in the Panhandle swamps did in fact nearly starve. Incidents related to this scene have already been discussed (chapter 1); the episode itself occurred shortly after part of the mopping-up contingent had gone to Baton Rouge to rejoin Jackson's army. Crockett's party had gotten as far east as the Apalachicola River and was preparing to march north into Alabama, so it is unlikely that the river mentioned below (David calls it the "Scamby") was actually the Escambia, which flows out of South Alabama and across the west end of the Panhandle into Pensacola Bay.

Among the characteristic idiomatic phrases appearing here is the classic "root hog or die," and scholars attribute its coinage to Crockett himself (*N*, p. 118, fn. 6).

> As the army marched, I hunted every day, and would kill every hawk, bird, and squirrel that I could find. Others did the same; and it was a rule with us, that when we stop'd at night, the hunters would throw all they killed in a pile, and then we would make a general division among all the men. One evening I came in, having killed nothing that day. I had a very sick man in my mess, and I wanted something for him to eat, even if I starved myself. So I went to the fire of a Captain Cowen, who commanded my company after the promotion of Major Russell, and informed him that I was on the hunt of something for a sick man to eat. I know'd the captain was as bad off as the rest of us, but I

found him broiling a turkey's gizzard. He said he had divided the turkey out among the sick, that Major Smiley had killed it, and that nothing else had been killed that day. I immediately went to Smiley's fire, where I found him broiling another gizzard. I told him that it was the first turkey I had ever seen have two gizzards. But so it was, I got nothing for my sick man. And now seeing that every fellow must shift for himself, I determined that in the morning, I would come up missing; so I took my mess and cut out to go ahead of the army. We know'd that nothing more could happen to us if we went than if we stayed, for it looked like it was to be starvation any way; we therefore determined to go on the old saying, root hog or die. We passed two camps, at which our men that had gone on before us, had killed Indians. At one they had killed nine, and at the other three. About daylight we came to a small river, which I thought was the Scamby; but we continued on for three days, killing little or nothing to eat; till, at last, we all began to get nearly ready to give up the ghost, and lie down and die; for we had no prospect of provision, and we knew we couldn't go much further without it.

We came to a large prairie, that was about six miles across it, and in this I saw a trail which I know'd was made by bear, deer, and turkeys. We went on through it till we came to a large creek, and the low grounds were all set over with wild rye, looking as green as a wheat field. We here made a halt, unsaddled our horses, and turned them loose to graze.

One of my companions, a Mr. Van Zandt, and myself, then went up the low grounds to hunt. We had gone some distance, finding nothing, when at last, I found a squirrel, which I shot, but he got into a hole in the tree. The game was small, but necessity is not very particular; so I thought I must have him, and I climbed that tree thirty feet high, without a limb, and pulled him out of his hole. I shouldn't relate such small matters, only to show what lengths a hungry man will go to, to get something to eat. I soon killed two other squirrels, and fired at a large hawk. At this a large gang of turkeys rose from the canebrake, and flew across the creek to where my friend was, who had just before crossed it. He soon fired on a large gobbler, and I heard it fall. By this time my gun was loaded again, and I saw one sitting on my side of the creek, which had flew over when he fired; so I blazed away, and down I brought him. I gathered him up, and a fine turkey he was. I now began to think we had struck a breeze of luck, and almost forgot our past sufferings, in the prospect of once more having something to eat. I raised the shout, and my comrade came to me, and we went on to our camp with the game we had killed. While we were gone, two of our mess had been out, and each of them had found a bee tree. We turned into cooking some of our game, but we had neither salt nor

bread. Just at this moment, on looking down the creek, we saw our men who had gone on before us for provisions, coming to us. They came up, and measured out to each man a cupful of flour. With this, we thickened our soup, when our turkey was cooked, and our friends took dinner with us, and then went on.
—*A Narrative of the Life of David Crockett* (1834), chapter 8.

BEARS, WRECKS, AND CHAOS

The following three selections from Crockett's autobiography might well be classified as definitive examples of backwoods humor. They match nicely the raw, rugged stories associated with "humor of the Old Southwest." In many ways, they are not really funny: the first hunting story depends solely upon its unusual action, which is chaotic and furious, while the story about Crockett's river disaster is humorous only because he made it so by his way of telling. But in the Old Southwest, hunting yarns and disaster stories, both true and invented, were often thought of as kinds of humor even if they weren't very humorous. People simply thought of them as belonging to a genre of popular humor, just as we refer to a certain Sunday supplement as the funny papers even though some of the features aren't really funny, and as we call certain magazines comic books even when they aren't comic. And, of course, his way of telling the story is the most crucial ingredient of all. Crockett was nearly killed in the wreck of his rafts, but not the slightest trace of anxiety remains in the story itself, for, after all, the primary purpose of humor is to transcend our ever-present awareness of death's immediacy. Good storyteller that he is, Crockett makes it look like a joke upon himself.

All three of these selections are rich with backwoods idioms, but an equally important feature displayed here is Crockett's fast-paced, realistic narration—a style that is exactly the opposite of the sentimental and ornate literary effusions of the popular hack novelists of the day.

In the first selection, Crockett is clearly working hard to be truthful about the bear he killed, and later he says it was only slightly smaller than one he killed during another hunt and was able to weigh. That other bear tipped the scales at 617 pounds, so these were truly large black bears. The second selection seems to contain quite a bit of pure invention, and the part about Crockett's climbing up a tree and sliding down to get warm was in fact extensively reprinted in newspapers and almanacs as a prime example of his exaggerated yarn spinning. For all its slapstick comedy, the third story, about losing his carefully prepared load of staves, is probably true, just as he tells it.

> In a little time my dogs started a large gang of old turkey gobblers, and I killed two of them, of the biggest sort. I shouldered them up, and

moved on, until I got through the harricane, when I was so tired that I laid my gobblers down to rest, as they were confounded heavy, and I was mighty tired. While I was resting, my old hound went to a log, and smelt it awhile, and then raised his eyes toward the sky, and cried out. Away he went, and my other dogs with him, and I shouldered up my turkeys again, and followed on as hard as I could drive. They were soon out of sight, and in a very little time I heard them begin to bark. When I got to them, they were barking up a tree, but there was no game there. I concluded it had been a turkey, and that it had flew away.

When they saw me coming, away they went again; and, after a little time, began to bark as before. When I got near them, I found they were barking up the wrong tree again, as there was no game there. They served me in this way three or four times, until I was so infernal mad, that I determined, if I could get near enough, to shoot the old hound at least. With this intention I pushed on the harder, till I came to the edge of an open prairie, and looking on before my dogs, I saw in and about the biggest bear that ever was seen in America. He looked, at the distance he was from me, like a large black bull. My dogs were afraid to attack him, and that was the reason they had stop'd so often, that I might overtake them. They were now almost up with him, and I took my gobblers from my back and hung them up in a sapling, and broke like a quarter horse after my bear, for the sight of him had put new springs in me. I soon got near to them, but they were just getting into a roaring thicket, and so I couldn't run through it, but had to pick my way along, and had close work even at that.

In a little time I saw the bear climbing up a large black oak-tree, and I crawled on till I got within about eighty yards of him. He was setting with his breast to me; and so I put fresh priming in my gun, and fired at him. At this he raised one of his paws and snorted loudly. I loaded again as quick as I could, and fired as near the same place in his breast as possible. At the crack of my gun here he came tumbling down; and the moment he touched the ground, I heard one of my best dogs cry out. I took my tomahawk in one hand, and my big butcher-knife in the other, and run up within four or five paces of him, at which he let my dog go, and fixed his eyes on me. I got back in all sorts of a hurry, for I know'd if he got hold of me, he would hug me altogether too close for comfort. I went to my gun and hastily loaded her again, and shot him the third time, which killed him good.

—*A Narrative of the Life of David Crockett* (1834), chapter 12.

In the morning I left my son at the camp, and we started on towards the harricane; and when we had went about a mile, we started a very large bear, but we got along mighty slow on account of the cracks in the earth occasioned by the earthquakes. We, however, made out to keep in hearing of the dogs for about three miles, and then we come to the harricane. Here we had to quit our horses, as old Nick himself couldn't have got through it without sneaking it along in the form that he put on to make a fool of our old grandmother Eve. By this time several of my dogs had got tired and come back; but we went ahead on foot for some little time in the harricane, when we met a bear coming straight to us, and not more than twenty or thirty yards off. I started my tired dogs after him, and McDaniel pursued them, and I went on to where my other dogs were. I had seen the track of the bear they were after, and I know'd he was a screamer. I followed on to about the middle of the harricane; but my dogs pursued him so close, that they made him climb an old stump about twenty feet high. I got in shooting distance of him and fired, but I was all over in such a flutter from fatigue and running, that I couldn't hold steady; but, however, I broke his shoulder, and he fell. I run up and loaded my gun as quick as possible, and shot him again and killed him. When I went to take out my knife to butcher him, I found I had lost it in coming through the harricane. The vines and briers was so thick that I would sometimes have to get down and crawl like a varment to get through at all; and a vine had, as I supposed, caught in the handle and pulled it out. While I was standing and studying what to do, my friend came to me. He had followed my trail through the harricane, and had found my knife, which was mighty good news to me, as a hunter hates the worst in the world to lose a good dog, or any part of his hunting-tools. I now left McDaniel to butcher the bear, and I went after our horses, and brought them as near as the nature of the case would allow. I then took our bags, and went back to where he was; and when we had skin'd the bear, we fleeced off the fat and carried it to our horses at several loads. We then packed it up on our horses, and had a heavy pack of it on each one. We now started and went on till about sunset, when I concluded we must be near our camp; so I hollered and my son answered me, and we moved on in that direction to the camp. We had gone but a little way when I heard my dogs make a warm start again; and I jumped down from my horse and gave him up to my friend, and told him I would follow them. He went on to the camp, and I went ahead after my dogs with all my might for a considerable distance, till at last night came on. The woods were very rough and hilly, and all covered over with cane.

I now was compelled to move on more slowly; and was frequently falling over logs, and into the cracks made by the earthquakes, so that I was very much afraid I would break my gun. However, I went on about three miles, when I came to a good big creek, which I waded. It was very cold, and the creek was about knee-deep; but I felt no great inconvenience from it just then, as I was all over wet with sweat from running, and I felt hot enough. After I got over this creek and out of the cane, which was very thick on all our creeks, I listened for my dogs. I found they had either treed or brought the bear to a stop, as they continued barking in the same place. I pushed on as near in the direction to the noise as I could, till I found the hill was too steep for me to climb, and so I backed and went down the creek some distance till I came to a hollow, and then took up that, till I come to a place where I could climb up the hill. It was mighty dark, and was difficult to see my way or any thing else. When I got up the hill, I found I had passed the dogs; and so I turned and went to them. I found, when I got there, they had treed the bear in a large forked poplar, and it was setting in the fork.

I could see the lump, but not plain enough to shoot with any certainty, as there was no moonlight; and so I set in to hunting for some dry brush to make me a light; but I could find none, though I could find that the ground was torn mightily to pieces by the cracks.

At last I thought I could shoot by guess, and kill him; so I pointed as near the lump as I could, and fired away. But the bear didn't come; he only clomb up higher, and got out on a limb, which helped me to see him better. I now loaded up again and fired, but this time he didn't move at all. I commenced loading for a third fire, but the first thing I know'd, the bear was down among my dogs, and they were fighting all around me. I had my big butcher in my belt, and I had a pair of dressed buckskin breeches on. So I took out my knife, and stood, determined, if he should get hold of me, to defend myself in the best way I could. I stood there for some time, and could now and then see a white dog I had, but the rest of them, and the bear, which were dark colored, I couldn't see at all, it was so miserable dark. They still fought around me, and sometimes within three feet of me; but, at last, the bear got down into one of the cracks, that the earthquakes had made in the ground, about four feet deep, and I could tell the biting end of him by the hollering of my dogs. So I took my gun and pushed the muzzle of it about, till I thought I had it against the main part of his body, and fired; but it happened to be only the fleshy part of his foreleg. With this, he jumped out of the crack, and he and the dogs had another hard fight around me, as before. At last, however, they forced him back into the crack again, as he was when I had shot.

I had laid down my gun in the dark, and I now began to hunt for it; and, while hunting, I got hold of a pole, and I concluded I would punch him awhile with that. I did so, and when I would punch him, the dogs would jump in on him, when he would bite them badly, and they would jump out again. I concluded, as he would take punching so patiently, it might be that he would lie still enough for me to get down in the crack, and feel slowly along till I could find the right place to give him a dig with my butcher. So I got down, and my dogs got in before him and kept his head towards them, till I got along easily up to him; and placing my hand on his rump, felt for his shoulder, just behind which I intended to stick him . I made a lunge with my long knife, and fortunately stuck him right through the heart; at which he just sank down, and I crawled out in a hurry. In a little time my dogs all come out too, and seemed satisfied, which was the way they always had of telling me that they had finished him.

I suffered very much that night with cold, as my leather breeches, and every thing else I had on, was wet and frozen. But I managed to get my bear out of this crack after several hard trials, and so I butchered him, and laid down to try to sleep. But my fire was very bad, and I couldn't find any thing that would burn well to make it any better; and I concluded I should freeze, if I didn't warm myself in some way by exercise. So I got up, and hollered a while, and then I would just jump up and down with all my might, and throw myself into all sorts of motions. But all this wouldn't do; for my blood was now getting cold, and the chills coming all over me. I was so tired, too, that I could hardly walk; but I thought I would do the best I could to save my life, and then, if I died, nobody would be to blame. So I went to a tree about two feet through, and not a limb on it for thirty feet, and I would climb up it to the limbs, and then lock my arms together around it, and slide down to the bottom again. This would make the insides of my legs and arms feel mighty warm and good. I continued this till daylight in the morning, and how often I clomb up my tree and slid down I don't know, but I reckon at least a hundred times.

—*A Narrative of the Life of David Crockett* (1834), chapter 15.

Having now closed my hunting for that winter, I returned to my hands, who were engaged about my boats and staves, and made ready for a trip down the river. I had two boats and about thirty thousand staves, and so I loaded with them, and set out for New Orleans. I got out of the Obion river, in which I had loaded my boats, very well; but when I got into the Mississippi, I found all my hands were bad scared, and in fact I believe I was scared a little the worst of any; for I had never been down the river, and I soon discovered that my pilot was as ignorant of the business as myself. I hadn't gone far before I determined to lash the two boats together; we did so, but it made them so heavy and obstinate, that it was next akin to impossible to do any thing at all with them, or to guide them right in the river.

That evening we fell in company with some Ohio boats; and about night we tried to land, but we could not. The Ohio men hollered to us to go on and run all night. We took their advice, though we had a good deal rather not; but we couldn't do any other way. In a short distance we got into what is called the *"Devil's Elbow"*; and if any place in the wide creation has its own proper name, I thought it was this. Here we had about the hardest work that I ever was engaged in, in my life, to keep out of danger; and even then we were in it all the while. We twice attempted to land at wood-yards, which we could see, but couldn't reach.

The people would run out with lights, and try to instruct us how to get to shore; but all in vain. Our boats were so heavy that we couldn't take them much any way, except the way they wanted to go, and just the way the current would carry them. At last we quit trying to land, and concluded just to go ahead as well as we could, for we found we couldn't do any better. Some time in the night I was down in the cabin of one of the boats, sitting by the fire, thinking on what a hobble we had got into, and how much better bear-hunting was on hard land, than floating along on the water, when a fellow had to go ahead whether he was exactly willing or not.

The hatchway into the cabin came slap down, right through the top of the boat; and it was the only way out except a small hole in the side, which we had used for putting our arms through to dip up water before we lashed the boats together.

We were now floating sideways, and the boat I was in was the hindmost as we went. All at once I heard the hands begin to run over the top of the boat in great confusion, and pull with all their might; and the first thing I know'd after this we went broadside full tilt against the head of an island where a large raft of drift timber had lodged. The nature of such a place would be, as every body knows, to suck the boats down, and turn them right under this raft; and the uppermost boat would, of course, be suck'd down and go under first. As soon as we struck, I bulged for my hatchway, as the boat was turning under sure enough. But when I got to it, the water was pouring through in a current as large as the hole would let it, and as strong as the weight of the river could force it. I found I couldn't get out here, for the boat was now turned down in such a way, that it was steeper than a house-top. I now thought of the hole in the side, and made my way in a hurry for that. With difficulty I got to it, and when I got there, I found it was too small for me to get out by my own dower, and I began to think that I was in a worse box than ever. But I put my arms through and hollered as loud as I could roar, as the boat I was in hadn't yet quite filled with water up to my head, and the hands who were next to the raft, seeing my arms out, and hearing me holler, seized them, and began to pull. I told them I was sinking, and to pull my arms off, or force me through, for now I know'd well enough it was neck or nothing, come out or sink.

By a violent effort they jerked me through; but I was in a pretty pickle when I got through. I had been sitting without any clothing over my shirt: this was torn off, and I was literally skin'd like a rabbit. I was, however, well pleased to get out in any way, even without shirt or hide, as before I could straighten myself on the boat next to the raft, the one they pull'd me out of went entirely under, and I have never seen it any more to this day. We all escaped on to the raft, where we were compelled to sit all night, about a mile from land on either side. Four of my company were bareheaded, and three barefooted; and of that number I was one. I reckon I looked like a pretty cracklin' ever to get to Congress!

We had now lost all our loading; and every particle of our clothing, except what little we had on; but over all this, while I was setting there, in the night, floating about on the drift, I felt happier and better off than I ever had in my life before, for I had just made such a marvelous escape, that I had forgot almost every thing else in that; and so I felt prime.

—*A Narrative of the Life of David Crockett* (1834), chapter 16.

TWO USEFUL COONSKINS

Two important variations of Crockett's "A Useful Coonskin" follow. The first is Mathew St. Clair Clarke's rather colorless recapitulation of a yarn Crockett had told him. We probably should assume that the story suffered some loss in transit from Crockett to Clarke to print. The second version is from Richard Penn Smith's *Crockett's Exploits*, having gotten into that jury-rigged and hasty production by way of a letter sent to Carey and Hart, as explained earlier in chapters 1 and 2. (The story is part of the first chapter of *Crockett's Exploits*; in the compilation of Crockett's *Narrative*, *Crockett's Tour*, and *Crockett's Exploits* that is usually titled *Life of David Crockett*, this chapter is numbered 23.)

The longer, fully developed version is one of the most famous Crockett stories. Its complex richness of characterization and detail mark it as being of a different order than Clarke's brief retelling. For this reason, too, a juxtaposition of the stories shows how Crockett's legendary image—and his awareness of his image—had expanded. In the version from *Crockett's Exploits*, Crockett amplifies his own public character while reflecting it back to the public in a yarn clearly prepared for a wide audience.

The second version of the yarn contains a rare idiomatic phrase that almost certainly came into West Tennessee dialect from one of the dialect regions of England, in the characteristic pattern by which a great many English dialect features were transplanted into American speech. Crockett says he walked over to the bar with all his cheering followers behind him, but walked away alone after they saw that he could not get credit for a bottle of rum. He says he left the clearing "in another guess sort" from the way he came in. Though it sounds awkward now, this phrase, often shortened to "in another guess" or "in another sort," meant simply "in a different way" (cf. Wright, p. 753; Mathews, p. 31).

His fondness for fun gave rise to many anecdotes; among others I have heard this, which I do not altogether believe: Colonel Crockett, while on an electioneering trip, fell in at a gathering, and it became necessary for him to treat the company. His finances were rather low, having but one 'coon skin about him; however, he pulled it out, slapped it down on the counter, and called for its value in whiskey. The merchant measured out the whiskey and threw the skin into the loft. The colonel, observing the logs very open, took out his ramrod, and, upon the merchant turning his back, twisted his 'coon skin out and pocketed it: when more whiskey was wanted, the same skin was pulled out, slapped upon the counter, and its value called for. This trick was played until they were all tired drinking.

—*Sketches and Eccentricities of Colonel David Crockett* (1833), chapter 5.

While on the subject of election matters, I will just relate a little anecdote about myself, which will show the people to the east, how we manage these things on the frontiers. It was when I first run for Congress; I was then in favor of the Hero, for he had chalked out his course so sleek in his letter to the Tennessee legislature, that, like Sam Patch, says I, "there can be no mistake in him," and so I went ahead. No one dreamt about the monster and the deposits at that time, and so, as I afterward found, many, like myself, were taken in by these fair promises, which were worth about as much as a flash in the pan when you have a fair shot at a fat bear.

But I am losing sight of my story. Well, I started off to the Cross Roads, dressed in my hunting shirt, and my rifle on my shoulder. Many of our constituents had assembled there to get a taste of the quality of the candidates at orating. Job Snelling, a gander-shanked Yankee, who had been caught somewhere about Plymouth Bay, and been shipped to the west with a cargo of codfish and rum, erected a large shanty, and set up shop for the occasion. A large posse of the voters had assembled before I arrived, and my opponent had already made considerable headway with his speechifying and his treating, when they spied me about a rifle shot from the camp, sauntering along as if I was not a party in the business. "There comes Crockett," cried one. "Let us hear the colonel," cried another, and so I mounted the stump that had been cut down for the occasion, and began to bushwhack in the most approved style.

I had not been up long before there was such an uproar in the crowd that I could not hear my own voice, and some of my constituents let me know, that they could not listen to me on such a dry subject as the welfare of the nation, until they had something to drink, and that I must treat them. Accordingly I jumped down from the rostrum, and led the way to the shanty, followed by my constituents, shouting, "Huzza for Crockett," and "Crockett for ever!"

When we entered the shanty, Job was busy dealing out his rum in a style that showed he was making a good day's work of it, and I called for a quart of the best, but the crooked crittur returned no other answer than by pointing to a board over the bar, on which he had chalked in large letters, "*Pay to-day and trust to-morrow.*" Now that idea brought me up all standing; it was a sort of cornering in which there was no back out, for ready money in the west, in those times, was the shyest thing in all natur, and it was most particularly shy with me on that occasion.

The voters, seeing my predicament, fell off to the other side, and I was left deserted and alone, as the Government will be, when he no longer has any offices to bestow. I saw, as plain as day, that the tide of popular opinion was against me, and that, unless I got some rum speedily, I should lose my election as sure as there are snakes in Virginny,—and it must be done soon, or even burnt brandy wouldn't save me. So I walked away from the shanty, but in another guess sort from the way I entered it, for on this occasion I had no train after me, and not a voice shouted, "Huzza for Crockett." Popularity sometimes depends on a very small matter indeed; in this particular it was worth a quart of New England rum, and no more.

Well, knowing that a crisis was at hand, I struck into the woods with my rifle on my shoulder, my best friend in time of need, and as good fortune would have it, I had not been out more than a quarter of an hour before I treed a fat coon, and in the pulling of a trigger, he lay dead at the root of the tree. I soon whipped his hairy jacket off his back, and again bent my steps towards the shanty, and walked up to the bar, but not alone, for this time I had half a dozen of my constituents at my heels. I threw down the coon skin upon the counter, and called for a quart, and Job, though busy in dealing out rum, forgot to point at his chalked rules and regulations, for he knew that a coon was as good a legal tender for a quart, in the west, as a New York shilling, any day in the year.

My constituents now flocked about me, and cried, "Huzza for Crockett," "Crockett for ever," and finding the tide had taken a turn, I told them several yarns, to get them in a good humor, and having soon dispatched the value of the coon, I went out and mounted the stump, without opposition, and a clear majority of the voters followed me to hear what I had to offer for the good of the nation. Before I was half through, one of my constituents moved that they would hear the balance of my speech, after they had washed down the first part with some more of Job Snelling's extract of cornstalk and molasses, and the question being put, it was carried unanimously. It wasn't considered necessary to tell the yeas and nays, so we adjourned to the shanty, and on the way I began to reckon that the fate of the nation pretty much depended upon my shooting another coon.

While standing at the bar, feeling sort of bashful while Job's rules and regulations stared me in the face, I cast down my eyes, and discovered one end of the coon skin sticking between the logs that supported the bar. Job had slung it there in the hurry of business. I gave it a sort of quick jerk, and it followed my hand as natural as if I had been the rightful owner. I slapped it on the counter, and Job, little dreaming that he was barking up the wrong tree, shoved along another bottle, which my constituents quickly disposed of with great good humor, for some of them saw the trick, and then we withdrew to the rostrum to discuss the affairs of the nation.

I don't know how it was, but the voters soon became dry again, and nothing would do, but we must adjourn to the shanty, and as luck would have it, the coon skin was still sticking between the logs, as if Job had flung it there on purpose to tempt me. I was not slow in raising it to the counter, the rum followed of course, and I wish I may be shot, if I didn't, before the day was over, get ten quarts for the same identical skin, and from a fellow, too, who in those parts was considered as sharp as a steel trap, and as bright as a pewter button.

This joke secured me my election, for it soon circulated like smoke among my constituents, and they allowed, with one accord, that the man who could get the whip hand of Job Snelling in fair trade, could outwit Old Nick himself, and was the real grit for them in Congress. Job was by no means popular; he boasted of always being wide awake, and that any one who could take him in, was free to do so, for he came from a stock, that sleeping or waking had always one eye open, and the other not more than half closed. The whole family were geniuses. His father was the inventor of wooden nutmegs, by which Job said he might have made a fortune, if he had only taken out a patent and kept the business in his own hands; his mother Patience manufactured the first white oak pumpkin seeds of the mammoth kind, and turned a pretty penny the first season; and his aunt Prudence was the first to discover that corn husks, steeped into tobacco water, would make as handsome Spanish wrappers as ever came from Havana, and that oak leaves would answer all the purpose of filling, for no one could discover the difference except the man who smoked them, and then it would be too late to make a stir about it. Job, himself, bragged of having made some useful discoveries, the most profitable of which was the art of converting mahogany sawdust into cayenne pepper, which he said was a profitable and safe business; for the people have been so long accustomed to having dust thrown in their eyes, that there wasn't much danger of being found out.

The way I got to the blind side of the Yankee merchant was pretty generally known before election day, and the result was, that my opponent might as well have whistled jigs to a milestone, as attempt to beat up for votes in that district. I beat him out and out, quite back into the old year, and there was scarce enough left of him, after the canvass was over, to make a small grease spot. He disappeared without even leaving a mark behind; and such will be the fate of Adam Huntsman, if there is a fair fight and no gouging.

After the election was over, I sent Snelling the price of the rum, but took good care to keep the fact from the knowledge of my constituents. Job refused the money, and sent me word, that it did him good to be taken in occasionally, as it served to brighten his ideas; but I afterwards learnt when he found out the trick that had been played upon him, he put all the rum I had ordered in his bill against my opponent, who, being elated with the speeches he had made on the affairs of the nation, could not descend to examine into the particulars of a bill of a vender of rum in the small way.

—*Col. Crockett's Exploits and Adventures in Texas* (1835), chapter 1.

WITH THE BARK OFF

In the Disney series, Fess Parker made a running joke out of Davy's legendary talent for grinning a raccoon out of a tree. It was also said that Davy's grin would pacify an angry bear. The tradition of Crockett's devastating grin had its beginning in the following passage from Clarke's *Sketches*. Clarke here gives us this yarn in the form of a stump speech by Crockett. For many years after the publication of *Sketches*, the almanacs and newspapers used the phrase "With the Bark Off" as a heading for any tall tale or humorously exaggerated anecdote. Crockett's grin and its astonishing effect had become a metaphor for humorous yarn spinning, which is, after all, a harmless and productive form of comic aggression.

Clarke tells us that Crockett's opponent in this particular election was a man who could influence voters by "wearing upon his countenance a peculiarly good-humored smile." Crockett opens his speech by referring to his opponent's famous political attribute.

> Yes, gentlemen, he may get some votes by *grinning*, for he can *out-grin* me, and you know I ain't slow—and to prove to you that I am not, I will tell you an anecdote. I was concerned myself—and I was fooled a little of the wickedest. You all know I love hunting. Well, I discovered a long time ago that a 'coon couldn't stand my grin. I could bring one tumbling down from the highest tree. I never wasted powder and lead, when I wanted one of the creatures. Well, as I was walking out one night, a few hundred yards from my house, looking

carelessly about me, I saw a 'coon planted upon one of the highest limbs of an old tree. The night was very *moony* and clear, and old Ratler was with me; but Ratler won't bark at a 'coon—he's a queer dog in that way. So, I thought I'd bring the lark down, in the usual way, *by a grin*. I set myself—and, after grinning at the 'coon a reasonable time, found that he didn't come down. I wondered what was the reason—and I took another steady grin at him. Still he was *there*. It made me a little mad; so I felt round and got an old limb about five feet long—and, planting one end upon the ground, I placed my chin upon the other, and took *a rest*. I then grinned my best for about five minutes—but the cursed 'coon hung on. So, finding I could not bring him down by grinning, I determined to have him—for I thought he must be a droll chap. I went over to the house, got my axe, returned to the tree, saw the 'coon still there, and began to cut away. Down it come, and I run forward; but d--n the 'coon was there to be seen. I found that what I had taken for one, was a large knot upon a branch of the tree—and, upon looking at it closely, I saw that *I had grinned all the bark off, and left the knot perfectly smooth.*

Now, fellow citizens, you must be convinced that, in the *grinning line*, I myself am not slow—yet, when I look upon my opponent's countenance, I must admit that he is my superior. You must all admit it. Therefore, be wide awake—look sharp—and do not let him grin you out of your votes.

—Sketches and Eccentricities of Colonel David Crockett (1833), chapter 10.

CROCKETT'S FIRST SPEECH IN CONGRESS

A strong tradition of backwoods bragging which depends upon language formulas was discussed in chapter 2. The following version of Crockett's bragging speech appeared in a Crockett almanac (see Meine, *Almanacks*, pp. 106-7; Botkin, p. 28). This piece is interesting for several reasons. It purports to show Crockett talking in his usual backwoods manner in the hallowed halls of Congress. It contains more than a trace of violence and ends in a vulgar racist slur of the kind that is more often found in the later almanacs.

Curiously, this speech was used by the Disney team as the basis for one given by Fess Parker in the second installment of the television series, "Davy Crockett Goes to Congress." Parker's speech starts, however, with a version of the half horse, half alligator formula, and he adds that he has a little touch of the snapping turtle in him, too. This detail derives from one of the versions Clarke included in *Sketches* (as quoted in the earlier discussion). Parker's speech then moves into the "surest rifle and ugliest dog" sentence of the following selection. Some of the middle part of the almanac speech is deleted from Parker's, as is, of course, the closing boast that he can "swallow a nigger whole." It is also intriguing to observe that in the same installment a variant of the closing line is included in a bragging performance by a man who is an enemy of Crockett's in backwoods Tennessee. The racist slur is there replaced by the words, "I've et better men than you—if they heads was buttered and they ears pinned back." Clearly, Disney's version of Crockett's image was cleaned up in an appropriate way.

> I say, Mr. Speaker, I've had a speech in soak this six months, and it has swelled me like a drowned horse; if I don't deliver it, I shall burst and smash the windows. The gentleman from Massachusetts talks of summing up the merits of the question, but I'll sum up my own. In one word I'm a screamer, and have got the roughest racking horse, the prettiest sister, the surest rifle and the ugliest dog in the district. I'm a leetle the savagest crittur you *ever did see*. My father can whip any man in Kentucky, and I can lick my father. I can outspeak any man on this floor, and give him two hours start. I can run faster, dive deeper, stay longer under, and come out drier, than any *chap* this side the big *Swamp*. I can outlook a panther and outstare a flash of lightning, tote a steamboat on my back and play at rough and tumble with a lion, and take an occasional kick from a *zebra*. To sum up all in one word, *I'm a horse*. Goliath was a pretty hard colt, but I could choke him. I can take the rag off—frighten the old folks—astonish the natives—and beat the Dutch all to smash—make nothing of sleeping under a blanket of snow—and don't mind being frozen more than a rotten apple.
>
> Congress allows *lemonade* to the members and has it charged under the head of stationery—I move also that *whiskey* be allowed under the item of *fuel*. For *bitters* I can suck away at a noggin of aquafortis, sweetened with brimstone, stirred with a lightning rod, and skimmed with a hurricane. I've soaked my head and shoulders in Salt River so much that I'm always corned. I can walk like an ox, run like a fox, swim like an eel, yell like an Indian, fight like the devil, spout like an earthquake, make love like a mad bull, and swallow a nigger whole without choking if you butter his head and pin his ears back.
>   —*Davy Crockett's Almanack* (1837).

THE DOG AND THE RACCOON

In chapter 2, the following two stories were compared for the purpose of demonstrating how an established story about some other person could become a Crockett story. When this happened, somebody did considerable rewriting, of course. The first version below appeared in a letter published in the New York *Spirit of the Times* in 1832, and was referred to for many years as "Captain Scott's Coon Story." The second version, much shortened and skillfully reconstructed into a recognizable backwoods yarn, appeared a decade later in a Crockett almanac (see Botkin, p. 25; Dorson, *Davy Crockett*, pp. 111-12; Blair and Hill, p. 128; Yates, p. 171). The detail about Crockett's sympathy and generosity—followed by the raccoon's quite sensible suspicion—nicely exemplifies the mixture of nobility and humor in the public's idea of the legendary Davy.

When the old rifle regiment was stationed at Fort Smith, Captain Scott, then Lieutenant Scott, was stationed at that post. He was, perhaps, a better shot at that time than he has ever been since, for since then he has received an injury in his right arm. I well know that it was very common for him at that time, on a misty day, to sit on the upper gallery or stoop of his quarters and shoot the common chimney swallow on the wing, with as unerring certainty as one of our backwoodsmen would hit the paper on a target at sixty yards at a beef shooting. At the same post was another officer, a Lieutenant Van Swearengen, who, though much addicted to the pleasure of hunting, was a notoriously bad shot.

It appears that a dog had treed a raccoon in a very tall cottonwood, and after barking loud and long to no purpose, the coon expostulated with him, and endeavored to convince him of the absurdity of spending his time and labor at the foot of the tree, and assured him that he had not the most distant idea of coming down the tree, and begged him as a fellow-creature to leave him to the enjoyment of his rights.

The dog replied naturally, but, I fear, not in the same conciliatory style as the coon, and threatened him with the advent of some one who would bring him down. At this moment a cracking in the cane indicated the approach of some individual; the coon asked the dog who it was. The dog replied, with some exultation, that it was Lieutenant Van Swearengen. The coon laughed, and he laughed with a strong expression of scorn about his mouth: "Lieutenant Van Swearengen, indeed—he may shoot and be d--ned."

Van Swearengen made five or six ineffectual shots, and left the coon, to the great discomfiture of the dog, still unscathed, and laughing on the top of the tree. The dog smothered his chagrin by barking louder and louder, and the coon laughed louder and louder, until the merriment of the one, and the mortification of the other, was arrested by the approach of some other person.

The coon inquired who it was; the dog answered that it was Scott. "Who?" asked the coon, evidently agitated. "Why, Martin Scott, by G-d!" The coon cried in the anguish of despair, that he was a *gone coon*, rolled up the whites of his eyes, folded his paws on his breast, and tumbled out of the tree at the mercy of the dog, without making the least struggle for that life which he had, but a few minutes before, so vauntingly declared and believed was in no kind of danger.

Moral: There is no elevation in life that will justify us in an unbecoming levity towards our inferiors.
—*Spirit of the Times*, October 13, 1832.

A Sensible Varmint

Almost every boddy that knows the forrest understands parfectly well that Davy Crockett never loses powder and ball, havin' ben brought up to b'lieve it a sin to throw away amminition, and that is the bennefit of a vartuous eddikation.

I war out in the forrest one arternoon, and had jist got to a place called the Grate Gap, when I seed a rakkoon setting all alone upon a tree. I klapped the breech of Brown Betty to my sholder, and war jist a-goin' to put a piece of led between his sholders, when he lifted one paw, and sez he, "Is your name Crockett?"

Sez I, "You are rite for wonst, my name is Davy Crockett."

"Then," sez he, "you needn't take no further trubble, for I may as well cum down without another word." And the cretur walked rite down from the tree, for he considered himself shot.

I stoops down and pats him on the head, and sez I, "I hope I may be shot myself before I hurt a hare of your head, for I never had sich a kompliment in my life."

"Seeing as how you say that," sez he, "I'll jist walk off for the present, not doubting your word a bit, d'ye see, but lest you should kinder happen to change your mind."
—*Ben Hardin's Crockett Almanac* (1842).

4
THE CROCKETT RECORD

The study of either Crockett's life or his legend begins with the reading of his autobiography, *A Narrative of the Life of David Crockett of the State of Tennessee* (1834). James Shackford's dissertation (1948) was the basis for Stanley Folmsbee's annotated and documented 1973 edition. This is readily available and seems to have become the standard, but the student will find an excellent introduction, a good basic bibliography, and some useful historical footnotes in Joseph J. Arpad's edition (1972), which is "edited for the modern reader" and was apparently prepared for publication at about the same time as Folmsbee's. Since the autobiography contains many of the facts of Crockett's life and displays the style which is the foundation of his legend, it can be considered the definitive map of the two regions which constitute Crockett country.

Shackford and Folmsbee's edition of the *Narrative* is a useful research tool for an interesting reason which goes beyond the fact that the book is based on Shackford's exhaustive investigation into the actual events of Crockett's life. The photographically reproduced pages of the first edition occupy one half of each page, and the footnotes stand to the right or left. The notes cite basic research materials that can be used to compare and contrast what Crockett wrote to what other people wrote about the same events. For instance, if the student wants to find accounts of the Creek Wars or Andrew Jackson's personality which might contradict Crockett's, he will find sources in the notes right next to Crockett's comments. The notes are weak, however, when they refer to Crockett's storytelling style. Neither Shackford nor Folmsbee appears to have known much about linguistic descriptions of dialects, and neither seems to have been aware that a literary text recapitulates speech by representative formulas rather than literal phonetic transcriptions. The notes to this edition say that the *Narrative* fails to display authentic backwoods language, but this is simply because the editors did not realize that when he wrote, Crockett represented backwoods dialect in an artful way by using idioms selectively.

Shackford's biography, *David Crockett: The Man and the Legend* (1956), and Dan Kilgore's account of Crockett's death, *How Did Davy Die?* (1978), are the two documents that are absolutely indispensable to further pursuit of the actual man's history. Both books are useful as guides to the sources for Crockett research in existence prior to their dates of publication, including unpublished documents. Kilgore understands that storytelling has its own kind of truth, but many earlier historians, including Folmsbee and Shackford, felt that the legendary Davy is a deceptive fabrication which gets in the way of serious fact finding. The student who is not so fiercely dedicated to the literal facts will instantly perceive that Shackford's work, like the work of all Crockett scholars—folklorists, literary historians, sociologists, and experts in popular culture, as well as historians—is motivated primarily by an interest in the fame of the legendary Davy, not the historical David.

Perhaps because the legend outweighs the facts, the purposes of historical research into Crockett's life have not changed much during the last three or four decades. Before the publication of Shackford's biography, investigators who were not as exact or patient as Shackford were commonly misled by the legendary materials, but their basic intent was to fix dates and establish facts. Since then, historians seem to have been committed to filling in the gaps and refining the accuracy of the biographical story. However, Kilgore's fine monograph is a genuinely important historical contribution. Immediately following the posthumous publication of Shackford's biography, Stanley Folmsbee and Anna Grace Catron added new material about Crockett's life and set straight several details in a monograph published as three successive articles in *East Tennessee Historical Society's Publications* for 1956, 1957, and 1958. These contain exhaustive footnotes that would be valuable to any scholar, but the journal is not available in all libraries.

The student researching the facts of Crockett's life is well advised to use great care when stepping out of the lines established by Shackford, Folmsbee and Catron, Kilgore, and Arpad. As discussed in chapter 2, any "autobiography" or "life" of Crockett up to and including Hamlin Garland's *Autobiography of Crockett* (1923) is nothing more than a compilation or paraphrase of Crockett's *Narrative* and two of the spurious Crockett books. A few of these might be useful to the historian, but are probably of more interest to the student of popular culture: Edward Ellis's *Life* (1884) and Charles Allen's *David Crockett* (1911) are entertaining in their own right, as is William F. Cody's section on Crockett in his illustrated *Story of the Wild West* (1888), which also has sections on Daniel Boone, Kit

Carson, and Buffalo Bill himself. Emerson Hough (1903) writes with some accuracy about Crockett, but John S. C. Abbott's paraphrase (1874) is merely silly. Of course, Garland's edition remains useful because of its introduction.

A biographer writing about the life of Crockett is obviously not involved in the investigation of a man whose political actions had crucial historical importance. Instead we are interested in Crockett's public image and the processes by which his legend is created and re-created. The scholar who plans to write the biography of an ordinary person in order to reveal something about everyday life during a particular era would certainly find less distorted sources elsewhere, in the diaries and albums of people who had no flirtation with greatness. It thus seems unlikely that open-minded students will devalue Constance Rourke's *Davy Crockett* (1934) simply because Shackford and Folmsbee would have us do so. Rourke retells the legend's central stories and then incorporates other old yarns to capture a strong sense of Crockett's world. Moreover, her book does contain a good amount of judiciously employed historical fact, nicely woven into her rich evocation of time and place.

Open-minded students will also want to include as one of their first sources Walter Blair's article on the six-sided popular image of Crockett (1940); his fictionalized *Davy Crockett—Frontier Hero* (1955), which contains the most accessible account of Crockett's Indian removal speech; Richard Dorson's (1939) and Franklin Meine's (1955) collections of almanac yarns; and John Seelye's account of the curious history of the Nashville almanac hoax (1980). Further, Blair's article shows us that the facts behind a legend cannot be as interesting as the variations created by the imagination of a nation. Collectively, people expand and transform historical details into legends, and the products of this process reveal their most cherished values and their darkest fears. In the case of the Crockett legend, the process also reveals what Americans have found ironic and humorous about their history and their stereotypes.

However, when looking for commentaries on Crockett in the older syntheses of American intellectual history, the student should be prepared to encounter the scorn of the fact-finding historian. V. L. Parrington (1927), Russell Blankenship (1931), and Charles and Mary Beard (1927) are classic examples; these serious observers saw Crockett as a loutish democrat, the worst example of a horde of illiterate westerners who shoved their way first into state politics and then, during Jackson's presidency, into Washington (Parrington, pp. 172-79; Blankenship, pp. 223-25; Beard, pp. 534-41). More balanced traditional assessments were offered by critics who wrote about the literary history of the South: Samuel Link (1899) saw that Crockett's own humorous style of yarn spinning contributed to the interesting blend of the actual and the fictive in his public image (pp. 534-40); Jay Hubbell (1954) fit Crockett into the historical context of tall-tale America (pp. 662-66).

Perceptions like Link's and Hubbell's characterize the important work on Crockett by those scholars interested in American humor whose contributions are discussed throughout this volume: Walter Blair, Benjamin A. Botkin, V. L. O. Chittick, Bernard DeVoto, Richard Dorson, J. A. Leo Lemay, Franklin J. Meine, Constance Rourke, and John Seelye. Likewise, the several "truthful fictions" of the 1930s, 1940s, and 1950s—Blair's, Rourke's, Dee Brown's, Meridel Le Sueur's, Edwin Mayer's, Irwin Shapiro's, and Stewart Holbrook's—are not to be dismissed as pseudohistories, because they are consciously and conscientiously intended to be fictive extensions of the comic legend. The researcher should take special note of Joseph Arpad's article on Crockett, John Wesley Jarvis, and James Kirke Paulding (1965), and his study of the history of the fight story (1973), which shows how Crockett's image came to be associated with frontier bragging. Guy S. Miles (1956) and Lowell Harrison (1969) have written articles about the evolution of the comic Crockett, and Kenneth Porter (1943) discusses the evolution of the story of the useful coonskin—an article of great importance to Dorson's analysis of Crockett folklore in *America in Legend* (1973). As this volume was being written, the *Mississippi Folklore Register* published a superb article by Michael A. Lofaro (1980) on the role of Crockett and Daniel Boone in the history of frontier humor. The student interested in Crockett's humorous image would do well to begin with Lofaro's essay.

Noteworthy advances in Crockett scholarship have been made by people interested, as were Rourke and Blair, in new ways to perceive the symbolic value of the legend. Such work is interdisciplinary and often combines historical research with sociological theory, mythic criticism, or literary analysis. It is devoted to defining the types of popular art forms in which Crockett's image appears and the inclinations of the audience that responds to these art forms. The basic assumption of such work is that the art form and the audience are inseparable: ideas held by large groups of people help form the fictive image of a legendary figure, and the fiction in turn shapes and expands their culture. Sometimes the real person from whom the popular image is derived is in turn affected. We have seen how Crockett reacted to meeting his fictive self on the stage in front of him, and we have speculated that his own perceptions of his fictive self may have influenced his political behavior and even the manner of his death.

Two important studies devoted strictly to Crockett's popular image are Joseph J. Arpad's and Margaret King's dissertations. Arpad's work (1970) stresses Crockett's image as "an Original Legendary Eccentricity and Early American Character." His bibliography is extensive and accurately lists important research up to the late 1960s. It serves especially well as a guide to assessments of the various popular ideas of Crockett. King's dissertation (1976) examines the Crockett craze precipitated by the Walt Disney television series. She is interested in why the legend appeals so strongly to the public imagination, and her notes and bibliography cite sociological books and articles that discuss the mechanisms and effects of hero worship. Anyone who wants a clear picture of the various controversies spawned by the Crockett craze needs to consult her bibliography, since it goes so far as to list even letters to the editors of periodicals like *Harper's, Saturday Review*, and *New Yorker*, and short articles in pictorial magazines such as *Look* and *Life*. Four excellent articles on the mythic Crockett as perceived by his American audience are M. J. Heale's discussion of Crockett, the frontier, and Jacksonian politics (1973); Stuart A. Stiffler's study of the genesis of Crockett's heroic myth (1957); and two essays (1978, 1979) by Catherine L. Albanese on Crockett, myth, nature, and religion. Albanese tends to exaggerate Crockett's influence on the American mind, but her work is quite original, and she is breaking interesting ground. Her articles are models for an entirely new kind of approach to the role of a specific figure in the intriguing processes of legend making.

Extending his or her interest in Crockett and the cultural context of his image, the student will want to read Richard Slotkin's monumental book *Regeneration through Violence: The Mythology of the American Frontier* (1973). Slotkin discusses Davy in several contexts and associates him with both the public image of Daniel Boone and the fictional hero of James Fenimore Cooper's Leatherstocking tales. Crockett exemplifies two important popular roles: the hunter-buffoon and the national hero as legendary Indian fighter. Most importantly,

> Two mythic associations made the Crockett figure acceptable as a symbol of American values—his association with the image of the self-restrained, professional hunter Boone through the invocation of hunting as his characteristic activity; and his association with the values of civilization, represented by the white woman and her male associate, the farmer, who together bring civilized value and progress into the wilderness. (p. 555)

Dixon Wecter placed Crockett in a trio that included Boone and, surprisingly, Johnny Appleseed; his book *The Hero in America* (1941) is, of course, an old standard (pp. 181-98). Daniel Boorstin has a fine chapter in *The Americans* (1965) titled "Heroes or Clowns? Comic Supermen from a Subliterature," and his discussion of the role of Crockett's public figure is nicely balanced (pp. 325-37). He judges Crockett to have been for some time the most widely known candidate for national hero worship, and he judges the Crockett almanacs—"fertilized by vulgar humor and the popular imagination"—to have been "one of the first characteristically American genres" (p. 330). The thoroughgoing student might also want to look at Marshall Fishwick's two books on American heroes (1954, pp. 70-72; 1969, pp. 70-83). Paul Sann's *Fads and Delusions of the American People* (1967) is a picture book supported by short essays, but his classification of

Crockett—and Crockett's image as portrayed by Fess Parker—is intriguing: Davy is one of America's "Tinsel Heroes" (pp. 27-30).

In another classic synthesis, *Anti-intellectualism in American Life* (1966, pp. 145-71), Richard Hofstadter sees a strong role for Crockett in a peculiarly stubborn streak of American intellectual history (perhaps we ought to say Crockett had a role in American anti-intellectual history). John G. Cawelti writes about the growth of Crockett's fictional self in his important book *Apostles of the Self-made Man* (1965, pp. 39-76). Two indispensable essays (both appearing in books) are Jules Zanger's "The Frontiersman in Popular Fiction, 1820-60" (1967), and Robert Jacobs's *"Tobacco Road*: Lowlife and the Comic Tradition" (1980).

Shackford's dissertation (1948) contains an enormous annotated bibliography of nearly 150 pages. The first section of this bibliography locates and describes materials known to have been written by Crockett or attributed to him, including both printed work and manuscripts in Crockett's hand. Some unpublished documents not by Crockett but referring to him are also discussed. However, the researcher should begin his search for manuscripts with the citations under Crockett's name in J. Albert Robbins's *American Literary Manuscripts* (1977). Robbins's and Shackford's listings overlap in just a few instances, and each mentions holdings not indicated in the other. Because Robbins's listing is the more thorough by far, the researcher should use Shackford's to cross-check Robbins's.

Robbins indicates that at least one and as many as five letters in Crockett's hand can be found in holdings at twenty-four different libraries. Many of these collections have other kinds of unpublished documents referring to Crockett, as do four additional libraries that hold no letters in Crockett's hand. A check of one of Robbins's citations—the holdings in Indiana University's Lilly Library—reveals two letters in Crockett's hand and eight documents by members of his family, six of which refer to him. Robbins does not list documents by members of an author's family unless they refer to the author; therefore, his citation includes the two letters by David and the six letters referring to David, but not the remaining two letters written by David's descendants. The Crockett letters are interesting: in one (January 6, 1831) he says that quarrels between southern state governments and Washington might lead to dissolution of the Union; in the other (May 27, 1834) he calls Jackson the first American king. Such findings suggest that historians might still pursue one course of action leading to a reassessment of Crockett's historical role. As a vociferous member of the disgruntled opposition, Crockett wrote a number of similarly sarcastic observations. His letters might well be worth collecting, editing, annotating, and publishing.

Other information in Shackford's 1948 bibliography that is not obsolete is in his chronological arrangement of the older "lives" of Crockett, most of which are pseudobiographies; his description of reminiscences about

Crockett written by people who knew him; and his list of government records—county, state, and federal. Anyone interested in Crockett family genealogy should refer first to Shackford and Folmsbee's edition of the *Narrative* and not to Shackford's dissertation, for the edition (p. xvii) distinguishes the newer, more reliable genealogical work from older investigations that have proved somewhat inaccurate. However, the genealogy that Folmsbee cites, by Robert Torrence and Robert Whittenburg (1956), was privately printed and is now almost impossible to obtain. One copy is in the Historical Society of Pennsylvania (Philadelphia), and it is not available through interlibrary loan.

To update the information in the bibliographies and notes in the essential books, articles, and dissertations already discussed, the researcher will want to consult the annual bibliographies of southern literature in the spring issues of *Mississippi Quarterly*. Jerry T. Williams's checklist of southern literature (1978) is a conflation of the *Mississippi Quarterly* annual listings for the years 1968 to 1975. In *The Frontier Humorists* (1975), M. Thomas Inge has a short but useful checklist (pp. 314-15) that includes a few items not mentioned in *Mississippi Quarterly*. The very meticulous student might also search for Crockett articles and books that have been reviewed in *Journal of American History* (until 1964, this was titled *Mississippi Valley Historical Review*) and *Tennessee Historical Quarterly*; both journals have annual indexes. Annual bibliographies containing citations of work on Crockett as a folk figure appear in *Southern Folklore Quarterly* and *Journal of American Folklore* (in the latter, the bibliographies are in the form of separate annual supplements). Literary articles with Crockett as their main subject will be indexed in the quarterly issues of *American Literature*, but these are few and far between and will probably be listed in *Mississippi Quarterly* as well. The journals mentioned so far publish scholarly articles; since Crockett's legend is largely an aspect of popular culture, more general articles about him will appear in popular magazines, and these are indexed in *Readers' Guide to Periodical Literature*.

Only a few portraits of David Crockett exist. In his biography (1956), Professor Shackford discusses two sources of information about Crockett's appearance: personal memoirs by people who knew him, and paintings (*DC*, pp. 281-91). Shackford found evidence of the survival of no more than eight or nine portraits of Crockett (two different references he could not trace may have been to the same painting). A few of the paintings Shackford identifies are unreliable either because the artist, in romanticizing the hero, produced a mawkishly distorted representation, or because the painting has faded with time and is illegible. Of the reliable portraits still extant, four stand out as being exceptionally useful. The New York Historical Society owns a small watercolor by Anthony Lewis DeRose; it depicts Crockett's face as seen from the right front. The frontispiece of Shackford's biography is a black and white reproduction of this DeRose portrait. A bust portrait of Crockett that is often published is one painted by Samuel Stillman Osgood. Shackford could not locate the original Osgood painting; but an apparently accurate engraving was made of it, and this reproduction comes close to being a standard portrait (see the illustrations for this volume). John Gadsby Chapman painted two very interesting pictures of Crockett, as discussed in chapter 2. One is a bust portrait now on display in the Alamo museum. The other is a full-length oil portrait about twenty-four inches high. This Chapman painting (see illustrations) may very well be the most important portrait of Crockett still in existence.

In his article on Chapman's notebook description of Crockett (1960), Curtis Carroll Davis mistakenly says the Chapman full-length portrait was lost in the fire that destroyed the Texas state capitol in Austin in 1881. Shackford learned from Crockett's grandson that the painting lost in the capitol fire was actually a portrait of Crockett sitting at a table in Congress (letter dated November 1947 to Shackford from Ashley Crockett, *DC*, pp. 288-89). Information (1982) sent this author by Cassandra LaSalandra, research associate at the Humanities Research Center in Austin, indicates that the Chapman full-length portrait is in the Academic Center of the University of Texas, where it is part of the J. Frank Dobie collection.

The Chapman full-length portrait contradicts the stereotyped image of Crockett. He is clearly not tall and rangy; he is of medium height and sturdy build. His hunter's garb is realistic, and does not include the legendary fringed jacket and raccoon or wildcat-skin cap. Could it be that this Chapman portrait depicts the real David Crockett?

Travelers as well as students may want to find information about other important Crockett artifacts and about the locations of the historical events of his life. As Shackford (1956) works his way across the Crockett map, he mentions several monuments marking important events in the man's life; the preface to his and Folmsbee's edition of the *Narrative* contains a few additional leads. Robert M. McBride's *More Landmarks of Tennessee History* (1969) contains information about Crockett memorials in Tennessee. Stewart Holbrook's *Davy Crockett* (1955) has an appendix listing and describing ten memorials to Crockett in Tennessee, two in Alabama, and three in Texas. Of these, the most important to the researcher of artifacts is the Alamo museum in San Antonio. Some interesting things can also be seen in the Crockett museum in Morristown, Tennessee, a place which represents the tavern Crockett's father kept when David was a boy. There is much confusion about Crockett's various rifles—"Old Betsy" and "Pretty Betsy" were the names of two of them—and this information is sorted out in Texas Jim Cooper's article, "A Study of Some David Crockett Firearms" (1966).

Every student of Crockett's legend immediately realizes that it is crucial to understand the spurious nature of many of the original stories. At the same time, he or she will want to remember that invention is the lifeblood of legend, and no Crockett story should ever be dismissed because it is not historically factual, or because it did not originate with Crockett, or because it cannot qualify as folklore. Our most important image of Crockett came originally from a work of art that meets none of these tests: James Kirke Paulding's play, *The Lion of the West* (written 1830, staged in 1831). James Tidwell's excellent modern edition (1954) serves as a guide to that image and to primary sources about the play and about James Hackett, the actor who created the role of Nimrod Wildfire.

Of the five books published over Crockett's name while he was alive or immediately after his death, only the *Narrative* is his, and it is well to keep in mind that he had some help from Thomas Chilton. On the other hand, Shackford may have given Chilton too much credit for his contribution to Crockett's book (see Shackford's 1951 article on the authorship, as well as his biography). In his edition, Joseph Arpad argues that the composition is Crockett's, while matters of grammar, spelling, and punctuation fell to Chilton (Arpad, *A Narrative*, pp. 26-27). Identifying the author of the first spurious Crockett book, *Sketches and Eccentricities of Colonel David Crockett of West Tennessee* (1833), may be a controversial matter. In his preface to the modern edition of the *Narrative*, Folmsbee says that Shackford's assignment of authorship of *Sketches* to Mathew St. Clair Clarke is "conclusive" (p. xii, fn. 14; see *DC*, pp. 258-64). Arpad (*A Narrative*, p. 23) has continued to assume that the author was James Strange French, but the main evidence for French's authorship is that he was named the copyright holder in the original book contract with Harper Brothers; this arrangement may have been a deception on the part of the collaborators who helped Clarke produce the book, as explained by Shackford (and as discussed in chapter 2). Both *The Life of Martin Van Buren* and *An Account of Colonel Crockett's Tour to the North and Down East* were Whig party productions published over Crockett's name in 1835. *Col. Crockett's Exploits and Adventures in Texas* (1836) does contain two chapters (the first two) that can be ascribed to Crockett; the rest, as we have seen, is to be chalked up to Richard Penn Smith and to the various authors whose writings he so freely plagiarized.

Until Shackford's biography was published in 1956, Crockett's *Narrative*, Clarke's *Sketches*, *Colonel Crockett's Tour*, and *Crockett's Exploits* provided the main sources for virtually all successive Crockett books, including most of the pseudobiographies. Even Hamlin Garland felt there were good reasons to include *Crockett's Exploits* as one of the Crockett books in his collection, in spite of the fact that he knew Crockett did not write it. It is difficult to obtain first or even early editions of any of the Crockett books, but it is easy to find the *Narrative, Colonel Crockett's Tour*, and *Crockett's Exploits* assembled in various combinations as a popular edition published thirty or more years later under some title claiming to be a "life" of Crockett. Most substantial libraries have several such "lives," and research libraries at major universities generally have microfilms of first editions of all five Crockett books written in the 1830s (the mock biography of Van Buren is the fifth). A photographic reprint of Clarke's *Sketches* was published by Arno Press (New York) in 1974. Any investigation or reinvestigation of the authorship of the five Crockett books should begin with Shackford's biography, which at present appears to offer an accurate account of their histories.

Rewarding research awaits the student who will further pursue the composition and text of Crockett's *Narrative*. Given Folmsbee's limitations as a literary critic and linguist, the language of Crockett's book remains unexamined. Especially productive would be a detailed investigation of Crockett's usages and formulaic phrases and the relationship of these idioms, first to idioms recorded elsewhere at the time, and second to idioms still in existence in American vernacular speech. In the introduction to his edition (*A Narrative*, pp. 29-30), Arpad discusses briefly the influence of Benjamin Franklin's autobiography on Crockett's and mentions Crockett's knowledge of Seba Smith's Major Jack Downing letters, which were circulating among the periodicals in the early 1830s. The latter were important examples of Down-East humor, and featured an unsophisticated bumpkin who made telling observations upon the hypocrisies of politics and society. Certainly much remains to be done toward finding out what else went into Crockett's *Narrative* and what kind of place it holds in literary history.

Constance Rourke's *Davy Crockett* (1934) contains the basic bibliographical description of most of the known Crockett almanacs. Her citations indicate that there are substantial collections of almanacs in the Library of Congress, the Carnegie Library in Nashville, and Franklin J. Meine's collection. The Meine collection is now housed in the Rare Book Room of the University of Illinois Library (Urbana campus), and it presently contains four Crockett almanacs. Her description itself derives from almanacs she saw in the library of the American Antiquarian Society. These are not now gathered in one collection; according to a letter to this author from Associate Librarian Frederick E. Bauer, Jr. (January 30, 1981), the Crockett almanacs in the American Antiquarian Society are catalogued by place of publication, date, and title. Two substantial collections of yarns from the almanacs are Richard M. Dorson's (1939) and Franklin Meine's (1955). Dorson's *America in Legend* is also important for information about how the yarns got into the almanacs, but there are a few inaccuracies in this book's footnotes. The original publishers of Dorson's collection of almanac yarns did not want to publish all the bibliographical details, so the information about the almanacs from which his selections are taken is in his article "The Sources of *Davy Crockett, American Comic Legend*" (1958). This summary by Dorson, Rourke's bibliographical essay in *Davy Crockett*, and Seelye's article (1980) are the basic guides to the almanacs at present. Michael Lofaro (Department of English, University of Tennessee) has generously contributed considerable information to this author about his current research into the almanacs. His work promises to be more thorough than either Rourke's or Dorson's, and Lofaro has already discovered at least one previously unlisted almanac.

Rourke also mentions three dime novels starring Crockett. Other titles can be found in Edward Leithead's article (1968) and in volume 1 of Albert Johannsen's book on the publishing house of Beadle and Adams (1950). The dime novels seem generally to have derived from the original Crockett books. Most of them were fragile and are extremely difficult to find in libraries now. They are often undated, but most seem to have been published in the late 1860s and throughout the 1870s. Edward Ellis's *Sockdolager!—A Tale of Davy Crockett* was reprinted in 1961; this book was drawn directly from the *Narrative*, *Colonel Crockett's Tour*, and *Crockett's Exploits*.

We are fortunate to be able to read Frank Murdock's *Be Sure You're Right* in volume 4 of Goldberg and Heffner's *America's Lost Plays* (1940). Walter Blair and Hamlin Hill discuss the significance of this long-running Crockett melodrama in *America's Humor* (1978), and anyone interested in the impact of the play should remember Hamlin Garland's introductory remarks to his Crockett book, as quoted and discussed in chapter 2. Excellent basic information about the production of Walt Disney's television series and movies (1954-1955) can be found in Christopher Finch's *Walt Disney* (1973). As discussed earlier, the sensational public reception of Disney's Davy Crockett is the subject of Margaret King's dissertation. The Disney Archives in Burbank, California, cannot lend materials, but the archivist, David R. Smith, has written this author (February 10, 1981) to indicate that information can be found there on the several re-releases of the television series. The archives' collection also includes half a dozen Disney-sponsored Crockett books.

As of the date of David Smith's letter, the men who sang the Disney Crockett ballad for the television series remain unidentified, and they are not credited on the copies of the sheet music seen by this author in the American music collection in the Lilly Library at Indiana University. Best-selling single records of the song were made in 1955 by Bill Hayes (Cadence) and Tennessee Ernie Ford (Capitol; see Shapiro, *Popular Music*, p. 134). The Lilly Library's wonderful collection also has a copy of "Go Ahead," commonly known in the 1830s as "Crockett's March"; it is subtitled "A March Dedicated to Colonel Crockett" and was copyrighted in 1835 by Firth and Hall of New York. B. A. Botkin's *A Treasury of American Folklore* (1944) has the words and music for both "The Ballad of Davy Crockett" (pp. 15-16) first published in 1834, and "Hunters of Kentucky" (pp. 9-12) as it was published in a collection of Woodworth's music in 1826. Joseph Arpad (*A Narrative*, pp. 33-34) mentions several curious Crockett artifacts which came out of the first Crockett fad—the one that occurred in the 1830s. One of these is *Crockett's Free-and-Easy Songbook* (1837), a mixed collection of stage songs and ballads. An important song written not long after the Disney fad of the 1950s is "Remember the Alamo" by Jane Bowers; it is sung by the Kingston Trio in an album titled *The Kingston Trio at Large* (Capitol Records T1199, no date). Bowers's song is interesting because it contains a line which refers to a "young Davy Crockett" (Crockett was fifty) who was "smilin' and laughin' " when he died, and because it shows the persistence into very modern times of the values generated by the Alamo drama.

A descriptive review of John Wayne's film *The Alamo* can be found in *Filmfacts* for November 18, 1960. Richard B. Dimmitt's *A Title Guide to the Talkies* (1965) mentions two other modern films in which Crockett is portrayed, and Leslie Halliwell's *The Filmgoer's Companion* (1970) identifies the actors: George Montgomery played Crockett in *Davy Crockett, Indian Scout* (Reliance Pictures, 1950), and Arthur Hunnicut played Crockett in *The Last Command* (Republic Pictures, 1955). *The Film Index* (1941) lists a 1910 silent picture about Crockett, and two issues of *Moving Picture World*, for April 23 and May 7, 1910, indicate in brief summaries that its title was *Davy Crockett*, the company which made it was Selig Polyscope, and the plot was taken from Murdock's play!

This chapter has discussed in some detail the methods and resources that constitute the foundation of formal research into Crockett's life and legend. If a person interested in Crockett just for the fun of it were to select an even dozen items, he or she would most certainly begin by reading the *Narrative*. Next in order would be James Shackford's authoritative biography, Dan Kilgore's monograph on Crockett's death, Constance Rourke's *Davy Crockett*, Franklin Meine's collection from the Nashville almanacs, and Dee Brown's novel, *Wave High the Banner*. If the student were then to add Walter Blair's article on the six Davy Crocketts, John Seelye's account of the almanac hoax, and Michael Lofaro's discussion of the roles of Crockett and Boone in American humor, he or she would be well on the way to understanding how the Crockett legend was made. The student's last book ought to place everything in a thoroughly illuminated social context, and Richard Slotkin's *Regeneration through Violence* would be an ideal choice. If, along the way, the opportunity would arise to see Fess Parker's and John Wayne's portrayals of Crockett (both the Disney series and *The Alamo* are frequently repeated on television), the student's impression would be a substantial one indeed.

5

THE CROCKETT CHRONOLOGY

1658   (Approximate date.) English ancestors of Crockett's mother, Rebecca Hawkins, arrive in Virginia.

1708   (Approximate date.) Joseph Louis Crockett and his wife, Sarah Stewart, migrate from Ireland to America. Crockett's great grandfather, William, is born either during the voyage or at some later date.

1716   (Approximate date.) Joseph Louis Crockett, William, and other family members move to Virginia.

1750   (Approximate date.) William Crockett and family move to North Carolina.

1771   The names of David Crockett (the grandfather) and two of his sons are written in the records at Lincolnton Court House, North Carolina.

1775   David Crockett (the grandfather), his wife, and their children join the great westward migration by crossing the Appalachians to settle in the territory that will become the western tip of North Carolina and the eastern tip of Tennessee.

1777   Crockett's grandparents are killed when a party of Creeks and Chickamaugas attack their homestead near Rogersville in Hawkins County, Tennessee. Crockett's father, John, is not at home. His Uncle Joseph is wounded; his Uncle James is captured, and remains a captive for twenty years.

1780   (Approximate date.) John Crockett marries Rebecca Hawkins, and they settle in the area that is to become Greene County, Tennessee.

——   In October, at Kings Mountain in North Carolina, John Crockett is one of the frontiersmen who defeat British troops under Patrick Ferguson, an action that significantly delays Cornwallis's northward march.

1786   David Crockett is born to John and Rebecca on August 17 in Greene County, Tennessee, near the mouth of Limestone Creek on the Nolichucky River.

1794   John Crockett's mill on Cove Creek in Greene County is washed out by a flood.

1796   John Crockett establishes a tavern on the Knoxville-Abingdon Road.

——   Tennessee is admitted to the Union.

1798   Young Crockett is hired out to Jacob Siler to help move cattle to Virginia. Siler refuses to release him, so the boy escapes and makes his way home through severe winter weather.

1799   Already thirteen, Crockett starts school, is punished by his father for playing hookey, and runs away.

1800   Crockett stands on a wharf in Baltimore and dreams of being a sailor. He goes to work for wagoners instead, and begins a long journey home.

1802   After crossing the New River in an ice storm, Crockett returns to his father's tavern for a tearful reunion with his family.

1805   Crockett plans to marry Margaret Elder, but she decides he is irresponsible and breaks their rather informal engagement.

1806   Crockett courts Mary (Polly) Finley, and they are married in Jefferson County, Tennessee, in August.

1811   In the fall, Crockett, Polly, and their small sons, John and William, move to Lincoln County in Tennessee and settle on Mulberry Creek, a tributary of the Elk River.

1812   In the Southeast, the war takes the form of conflict with both the British and the Indians, and settlers in Tennessee and Alabama prepare to defend their holdings.

1813   Crockett and his family move to Franklin County and establish a homestead on Bean's Creek, southwest of Winchester, Tennessee.

——   On August 30, Indians attack Fort Mims in southern Alabama, killing the 260 soldiers in the garrison and a like number of civilians.

——   In September, Crockett enlists with the militia unit forming in Winchester, Tennessee.

——   On November 3, Crockett participates in a massacre as Andrew Jackson's army wipes out the Indian town of Tallusahatchee.

——   By December 24, Crockett is home, having served his ninety-day enlistment, in spite of his later claim that he took part in a mutiny against Jackson, an incident that was actually nothing more than a brief confrontation staged by militiamen who wanted to go home after sixty days.

1814	In March, Jackson defeats the Creeks decisively in the Battle of Horseshoe Bend.
——	Crockett reenlists in September.
——	On November 7, Jackson wins a showdown with the Spanish authorities and British troops at Pensacola. Crockett arrives on November 8, in time to see the British fleet set sail from Pensacola Bay.
——	Crockett's adventures in the Florida swamps begin. He and his comrades chase Indians as far east as the Apalachicola River.
1815	Crockett's second enlistment is over in March, and he resigns at the rank of sergeant, a fact he will omit from the *Narrative* because he wants to be known as a common soldier.
——	In the summer, Polly Crockett dies. Crockett finds that he cannot manage his homestead and household alone.
1816	In either spring or summer, Crockett marries Elizabeth Patton.
——	While exploring new country, Crockett suffers his first bad attack of malaria. Elizabeth receives news that he is dead, and is astonished by his return.
1817	In the spring, the Crocketts move to Shoal Creek in Lawrence County, Tennessee. His neighbors informally choose Crockett to be a backwoods magistrate. In November, he is made justice of the peace by official state action.
1818	Crockett is elected colonel of the 57th Regiment of Militia in Lawrence County.
1821	Crockett is elected to the Tennessee state legislature, and begins his fight to protect the interests of settlers.
——	A flood destroys David and Elizabeth's mills and distillery on Shoal Creek.
1821-1822	During the winter, Crockett explores the Obion River country.
1822	The Crockett family moves to a homestead near the Obion.
1823	Crockett is reelected to the state legislature.
1826	In the spring, Crockett attempts to run two boats full of staves to New Orleans, and is wrecked in the Mississippi River. Having narrowly escaped drowning, he and his friends are taken to Memphis, where Crockett's national political career begins as a result of his meeting Marcus Winchester, who urges him to run for Congress.
1827-1829	Crockett is elected to the House of Representatives and serves in the Twentieth Congress.
1829-1837	Andrew Jackson is president of the United States.
1829-1831	Crockett is reelected and serves in the Twenty-first Congress.
1830	Crockett speaks out fiercely against the Indian removal bill, but

The Crockett Chronology 149

1831 later tries to suppress the record because his constituents do not approve of his stand.

1831 Crockett is defeated in his bid for reelection to Congress.

—— On April 25, James Kirke Paulding's play *The Lion of the West* opens at the Park Theater in New York City. James Hackett plays the hero, Nimrod Wildfire, and the public immediately identifies Wildfire as the very image of Crockett.

1833 Mathew St. Clair Clarke publishes *Life and Adventures of Colonel David Crockett of West Tennessee*, in Cincinnati. The book achieves great popularity under its otherwise identical second title, *Sketches and Eccentricities*, which is published in New York later in the year.

—— In March, *The Lion of the West* is produced at the Theatre Royal Covent Garden in London. It has a new title, *The Kentuckian, or A Trip to New York*. Later in the year, Hackett plays Wildfire in Washington, D.C., having made the play part of his repertory. Crockett goes to see him, and each bows to each in a monumental meeting of legends.

1833-1835 Crockett runs again for a seat in Congress, wins this time, and serves in the Twenty-third Congress.

1834 With the help of Thomas Chilton, Crockett writes and publishes his autobiography, *A Narrative of the Life of David Crockett of the State of Tennessee*.

—— Crockett makes a speaking tour to the East and New England. The Whigs are courting him as a possible presidential candidate, but his love of publicity interferes with his congressional duties.

1835 Crockett is defeated in his last bid for Congress.

—— Halley's comet appears, and it is said that Old Hickory has commissioned Crockett to wring its tail off.

—— Using Crockett's name as the author, the Whigs publish two books, one about his speaking tour and another that is a satirical biography of Martin Van Buren.

—— On November 1, Crockett and a few friends start out for Texas. Sometime during the winter, he explores the Red River valley.

1836 In January, Crockett is in San Augustine, Texas. He attends a political dinner in nearby Nacogdoches and tells the Texans that he has said to the voters of Tennessee, "You can go to hell; I'm going to Texas."

—— On March 6, the Texan fortress at the Alamo falls to Santa Anna's army. At about six o'clock in the morning, Crockett and half a dozen others are captured and brought before Santa Anna. As some staff officers attempt to intervene, Santa Anna orders the immediate execution of the prisoners. Crockett and his comrades are killed on the spot.

1836	Early in the summer, at Edward Carey's suggestion, Richard Penn Smith puts together *Col. Crockett's Exploits and Adventures in Texas,* "Written by Himself." In spite of having been almost wholly fabricated, the book is destined to be a cornerstone of the Crockett legend.
——	In the fall, Martin Van Buren is elected president of the United States.
1835-1856	Over forty-five Crockett almanacs are published in various cities by various publishers. (Nashville imprints from 1835 to 1838 appear to have been produced in Boston.)
1841	Crockett's son, John, obtains a land bill in Congress, succeeding where his father had failed.
1872-1896	Frank Murdock's play *Davy Crockett; Or, Be Sure You're Right, Then Go Ahead* runs continuously in the United States and England, with Frank Mayo in the hero's role.
1923	Hamlin Garland publishes *The Autobiography of David Crockett,* an edition including the *Narrative, Colonel Crockett's Tour, Crockett's Exploits,* and an introduction in which Garland recalls seeing Frank Mayo as Davy.
1934	Constance Rourke publishes *Davy Crockett,* a book that mixes fact and fiction to give an impression of the world from which the legend sprang.
1942	Dee Brown publishes *Wave High the Banner,* a novel based on Crockett's life and legend and set firmly among authentic historical details.
1954-1955	Walt Disney produces a Davy Crockett series for television and combines the segments into a motion picture, *Davy Crockett, King of the Wild Frontier.* A Davy Crockett fad sweeps the nation.
1956	James A. Shackford publishes a definitive biography of the historical David Crockett.
1960	John Wayne produces the film *The Alamo,* and takes a starring role as Crockett.
1973	Stanley J. Folmsbee completes and publishes James Shackford's edition of Crockett's autobiography, the *Narrative.*

BIBLIOGRAPHY

Abbott, John S. C. *David Crockett*. New York: Dodd, Mead, 1874.
Adams, Henry. *The Education of Henry Adams*. Edited by Ernest Samuels. Boston: Houghton Mifflin, 1974.
The Alamo. John Wayne, director. With John Wayne as Crockett, Laurence Harvey, and Richard Widmark. United Artists, 1960.
"*The Alamo*." A review. Filmfacts 3 (1960): 255-57.
Albanese, Catherine L. "Citizen Crockett: Myth, History, and Nature Religion." *Soundings: An Interdisciplinary Journal* 61 (1978): 87-104.
_____. "King Crockett: Nature and Civility on the American Frontier." *Proceedings of the American Antiquarian Society* 88 (1979): 225-49.
Allen, Charles Fletcher. *David Crockett, Scout, Small Boy, Pilgrim, Mountaineer, Soldier, Bear-Hunter, and Congressman, Defender of the Alamo*. Philadelphia: J. B. Lippincott, 1911.
American Comic Almanac. 1831-1838.
American Literature. 1929- .
Arpad, Joseph John. "David Crockett, An Original Legendary Eccentricity and Early American Character." Ph.D. dissertation, Duke University, 1970.
_____. "The Fight Story: Quotation and Originality in Native American Humor." *Journal of the Folklore Institute* 10 (1973): 141-72.
_____. "John Wesley Jarvis, James Kirke Paulding, and Colonel Nimrod Wildfire." *New York Folklore Quarterly* 21 (1965): 92-106.
_____, ed. *A Narrative of the Life of David Crockett of the State of Tennessee*. New Haven, Conn.: College & University Press, 1972.
Bauer, Frederick E., Jr. Letter (from the American Antiquarian Society Library) to the author. January 30, 1981.
Beard, Charles A., and Beard, Mary R. *The Rise of American Civilization*. New York: Macmillan, 1927. Vol. 1.
Ben Hardin's Crockett Almanac. See *Crockett Almanac*.

Bird, Robert Montgomery. *Nick of the Woods; or, the Jibbenainosay*. Edited by Curtis Dahl. New Haven, Conn.: College & University Press, 1967.

Blair, Walter. *Davy Crockett—Frontier Hero: The Truth as He Told It—The Legend as His Friends Built It*. New York: Coward-McCann, 1955.

―――. "Davy Crockett, Tennessee Settler" and "Davy Crockett, Soldier, Congressman and Comet Licker." In *Tall Tale America: A Legendary History of Our Humorous Heroes*. New York: Coward-McCann, 1944, pp. 66-94, 259.

―――. *Horse Sense in American Humor: From Benjamin Franklin to Ogden Nash*. Chicago: University of Chicago Press, 1942. Reprint. New York: Russell & Russell, 1962.

―――. "Six Davy Crocketts." *Southwest Review* 25 (1940): 443-62. Also incorporated into chapter 2 of Blair's *Horse Sense in American Humor*.

―――, and Hill, Hamlin. *America's Humor: From Poor Richard to Doonesbury*. New York: Oxford University Press, 1978.

―――, and Meine, Franklin J., eds. *Half Horse Half Alligator: The Growth of the Mike Fink Legend*. Chicago: University of Chicago Press, 1956. Reprint. New York: Arno Press, 1977.

Blankenship, Russell. *American Literature as an Expression of the National Mind*. New York: Henry Holt, 1931. Rev. ed., 1949.

Boorstin, Daniel J. *The Americans: The National Experience*. New York: Random House, 1965.

Botkin, Benjamin A., ed. *A Treasury of American Folklore*. New York: Crown Publishers, 1944.

Bowers, Jane. "Remember the Alamo." Sung by the Kingston Trio on *The Kingston Trio at Large*. Capitol, T1199, n.d.

Brown, Dee. *Bury My Heart at Wounded Knee*. New York: Holt, Rinehart & Winston, 1970.

―――. *Wave High the Banner: A Novel Based on the Life of Davy Crockett*. Philadelphia: Macrae-Smith, 1942.

Caruthers, William Alexander. *The Kentuckian in New York; Or, The Adventures of Three Southerns*. New York: Harper & Bros., 1834. Reprint. Ridgewood, N.J.: Gregg Press, 1968.

Cawelti, John G. *Apostles of the Self-made Man*. Chicago: University of Chicago Press, 1965.

Chittick, V.L.O., ed. *Ring-Tailed Roarers: Tall Tales of the American Frontier, 1830-60*. Caldwell, Idaho: Caxton Printers, 1941.

Clark, William (probable author). *An Account of Colonel Crockett's Tour*. See "Crockett, David."

Clarke, Mathew St. Clair (probable author). *Sketches and Eccentricities of Colonel David Crockett of West Tennessee* (usual title). New York: J. & J. Harper, 1833. Reprint of *Life and Adventures of Colonel David Crockett of West Tennessee*. Cincinnati: "For the Proprietor," 1833.

Clayton, Augustin Smith (probable author). *The Life of Martin Van Buren*. See "Crockett, David."

Clemens, Samuel Langhorne. *Adventures of Huckleberry Finn*, edited by Sculley Bradley et al. New York: W. W. Norton, 1961.

Cody, William F. *Story of the Wild West and Camp-Fire Chats, by Buffalo Bill,*

(Hon. W. F. Cody): *A Full and Complete History of the Renowned Pioneer Quartette, Boone, Crockett, Carson and Buffalo Bill*. Philadelphia: Historical Publishing Co., 1888.

Cohen, Hennig, and Dillingham, William, eds. *Humor of the Old Southwest*. Boston: Houghton Mifflin, 1964. 2nd ed. Athens: University of Georgia Press, 1975.

Cooper, Texas Jim. "A Study of Some David Crockett Firearms." *East Tennessee Historical Society's Publications* 38 (1966): 62-69.

Crockett Almanac. 1835-1856. Title, publisher, frequency of publication, and place of publication vary. About fifty separate almanacs are extant.

Crockett, David. *The Adventures of Davy Crockett, Told Mostly by Himself*. New York: Charles Scribner's Sons, 1934. A paraphrase of Crockett's *Narrative* and *Crockett's Exploits*.

———. "Congressman Crockett's Speech on Indian Removal: A Report on the Remarks Made in The House of Representatives, May 19, 1830." In Walter Blair (see under), *Davy Crockett—Frontier Hero*, pp. 211-15. A paraphrase.

———. Letter to Daniel W. Pounds. January 6, 1831. Bloomington, Ind.: Lilly Library.

———. Letter to T. J. Dobings. May 27, 1834. Bloomington, Ind.: Lilly Library.

———. *Life of David Crockett, the Original Humorist and Irrepressible Backwoodsman*. New York: Lovell, Coryell & Co., no date. A compilation of Crockett's *Narrative*, *Colonel Crockett's Tour*, and *Crockett's Exploits*.

———. *The Life of David Crockett, the Original Humorist and Irrepressible Backwoodsman: An Autobiography*. New York: A. L. Burt, 1902. A compilation similar to the above.

———. *A Narrative of the Life of David Crockett of the State of Tennessee*. Philadelphia: Carey & Hart, 1834. Facsimile edition with annotations and introduction, edited by James A. Shackford and Stanley J. Folmsbee. Knoxville: University of Tennessee Press, 1973.

"Crockett, David" (pseudonym; probable author, William Clark). *An Account of Colonel Crockett's Tour to the North and Down East*. Philadelphia: Carey & Hart, 1835.

——— (pseudonym; author, Richard Penn Smith). *Col. Crockett's Exploits and Adventures in Texas, Written by Himself*. Philadelphia: "T. K. & P. G. Collins" (Carey & Hart), 1836.

——— (pseudonym; probable author, Augustin Smith Clayton). *The Life of Martin Van Buren, Hair-Apparent to the "Government," and the Appointed Successor of General Jackson*. Philadelphia: Robert White, 1835.

Crockett's Free-and-Easy Song Book: A New Collection of the Most Popular Stage Songs. Philadelphia: J. Kay, Jr. & Bro., 1837.

Davis, Curtis Carroll. "A Legend at Full-Length: Mr. Chapman Paints Colonel Crockett—and Tells about It." *Proceedings of the American Antiquarian Society* 69 (1960): 155-74.

Davy Crockett. A silent motion picture. Selig Polyscope, 1910.

Davy Crockett and the River Pirates. See Walt Disney Productions.

Davy Crockett, Indian Scout. With George Montgomery as Crockett. Reliance Pictures, 1950.

Davy Crockett, King of the Wild Frontier. See Walt Disney Productions.

Davy Crockett's Almanack. See *Crockett Almanac.*

DC. See Shackford, *David Crockett.*

DeVoto, Bernard. *Mark Twain's America.* Boston: Houghton Mifflin, 1932.

Dimmitt, Richard B. *A Title Guide to the Talkies.* New York: Scarecrow Press, 1965.

Disney, Walt. See Walt Disney Productions.

Dorson, Richard M. *America in Legend: Folklore from the Colonial Period to the Present.* New York: Random House, 1973.

———, ed. *Davy Crockett: American Comic Legend.* New York: Rockland Editions, 1939. Reprint. New York: Arno Press, 1977.

———. "The Sources of *Davy Crockett, American Comic Legend.*" *Midwest Folklore* 8 (1958): 143-49.

Ellis, Edward S. ("Harry Hazard"). *The Bear-Hunter; or, Davy Crockett as a Spy.* New York: Beadle & Adams, 1876.

——— ("Charles E. Lasalle"). *Col. Crockett, the Bear King.* New York: Beadle & Adams, 1886.

———. *The Life of Colonel David Crockett.* Philadelphia: Porter & Coates, 1884.

———. *Sockdolager! A Tale of Davy Crockett, in which the Old Tennessee Bear Hunter Meets up with the Constitution of the United States.* Richmond: Virginia Commission on Constitutional Government, 1961. Reprint of "dime novel" condensation of Ellis's *Life.*

———. ("Charles E. Lasalle"). *The Texan Trailer; or, Davy Crockett's Last Bear Hunt.* New York: Beadle & Adams, 1871.

Ellms, Charles. *The Pirates Own Book.* Boston: Samuel N. Dickinson, 1837.

Exman, Eugene. *The Brothers Harper, 1817-1853.* New York: Harper & Row, 1965.

The Film Index: A Bibliography. New York: Museum of Modern Art Film Library and H. W. Wilson, 1941. Reprint. New York: Arno Press, 1966. Vol. 1.

Finch, Christopher. *Walt Disney: From Mickey Mouse to the Magic Kingdoms.* New York: Harry N. Abrams, 1973.

Fishwick, Marshall W. *American Heroes: Myth and Reality.* Washington, D.C.: Public Affairs Press, 1954.

———. *The Hero, American Style.* New York: David McKay, 1969.

Folmsbee, Stanley J., and Catron, Anna Grace. "The Early Career of David Crockett." *East Tennessee Historical Society's Publications* 28 (1956): 58-85.

———. "David Crockett: Congressman." *East Tennessee Historical Society's Publications* 29 (1957): 40-78.

———. "David Crockett in Texas." *East Tennessee Historical Society's Publications* 30 (1958): 48-74.

Ford, Tennessee Ernie. "The Ballad of Davy Crockett." Capitol Records, 1955. See also Walt Disney Productions.

Franklin, Benjamin. *Autobiography.* Edited by Leonard W. Labaree et al. New Haven: Yale University Press, 1964.

French, James Strange. *Elkswatawa; Or, The Prophet of the West: A Tale of the Frontier.* New York: Harper & Bros., 1836.

Garland, Hamlin, ed. *The Autobiography of David Crockett.* New York: Charles Scribner's Sons, 1923.

" 'Go Ahead' A March Dedicated to Colonel Crockett." Sheet music. New York: Firth and Hall, 1835.

Goldberg, Isaac, and Heffner, Hubert, eds. *America's Lost Plays*. Princeton, N.J.: Princeton University Press, 1940.
Halliwell, Leslie. *The Filmgoer's Companion*. 3rd ed. New York: Hill & Wang, 1970.
Harrison, Lowell H. "Davy Crockett: The Making of a Folk Hero." *Kentucky Folklore Record* 15 (1969): 87-90.
Hayes, Bill. "The Ballad of Davy Crockett." Cadence Records, 1955. See also Walt Disney Productions.
"Hazard, Harry" (pseudonym). See Ellis, Edward S.
Heale, M. J. "The Role of the Frontier in Jacksonian Politics: David Crockett and the Myth of the Self-Made Man." *Western Historical Quarterly* 4 (1973): 405-23.
Hofstadter, Richard. *Anti-intellectualism in American Life*. New York: Alfred A. Knopf, 1966.
Holbrook, Stewart H. *Davy Crockett: From the Backwoods of Tennessee to the Alamo*. New York: Random House, 1955.
Hough, Emerson. *The Way to the West, and the Lives of Three Early Americans: Boone—Crockett—Carson*. Indianapolis: Bobbs-Merrill, 1903.
Hubbell, Jay B. *The South in American Literature, 1607-1900*. Durham, N.C.: Duke University Press, 1954.
Inge, M. Thomas. *The Frontier Humorists: Critical Views*. Hamden, Conn.: Archon Books, 1975.
Irving, Washington. *A History of New York*. Edited by Edwin T. Bowden. New York: Twayne Publishers, 1964.
Jacobs, Robert D. "*Tobacco Road*: Lowlife and the Comic Tradition." In *The American South: Portrait of a Culture*, edited by Louis D. Rubin, Jr. Baton Rouge: Louisiana State University Press, 1980, pp. 206-26.
Johannsen, Albert. *The House of Beadle and Adams and Its Dime and Nickel Novels: The Story of a Vanished Literature*. Norman: University of Oklahoma Press, 1950. Vol. 1.
Johnston, Arch C. "A Major Earthquake Zone on the Mississippi." *Scientific American* 246 (April, 1982): 60-68.
Journal of American Folklore. 1888- .
Journal of American History. 1964- . Formerly *Mississippi Valley Historical Review*. 1914-1964.
Kilgore, Dan. *How Did Davy Die?* College Station: Texas A&M University Press, 1978.
King, Margaret Jane. "The Davy Crockett Craze: A Case Study in Popular Culture." Ph.D. dissertation, University of Hawaii, 1976.
La Fay, Howard. "Texas!" *National Geographic* 157 (1980): 440-83.
LaSalandra, Cassandra. Letters to the author with information sheet from the iconography collection regarding John Gadsby Chapman portrait of Crockett (Humanities Research Center, University of Texas). January 15, 1982; February 2, 1982.
"Lasalle, Charles E." (pseudonym). See Ellis, Edward S.
The Last Command. With Arthur Hunnicut as Crockett. Republic Pictures, 1955.
Leithead, Edward J. "Legendary Heroes and the Dime Novel." *American Book Collector* 18, 7 (1968): 22-27.

Lemay, J. A. Leo. "The Text, Tradition, and Themes of 'The Big Bear of Arkansas.'" *American Literature* 47 (1975): 321-42.
Le Sueur, Meridel. *Chanticleer of the Wilderness Road: A Story of Davy Crockett.* New York: Alfred A. Knopf, 1951.
Link, Samuel Albert. *Pioneers of Southern Literature.* Nashville, Tenn.: M. E. Church, 1899. Vol. 2.
Lofaro, Michael A. "From Boone to Crockett: The Beginnings of Frontier Humor." *Mississippi Folklore Register* 14 (1980): 57-74.
Longstreet, Augustus Baldwin. *Georgia Scenes.* Augusta, Georgia: The *State Rights Sentinel* Office, 1835.
McBride, Robert M. "David Crockett and His Memorials in Tennessee." *Tennessee Historical Quarterly* 26 (1967): 219-39. Reprinted in *More Landmarks of Tennessee History.* Nashville: Tennessee Historical Society and Tennessee Historical Commission, 1969.
Mathews, Mitford M., ed. *A Dictionary of Americanisms on Historical Principles.* Chicago: University of Chicago Press, 1951.
Mayer, Edwin Justus. *Sunrise in My Pocket, Or The Last Days of Davy Crockett: An American Saga.* New York: Julian Messner, 1941.
Meadowcroft, Enid L. *The Story of Davy Crockett.* New York: Grosset & Dunlap, 1952.
Meine, Franklin J., ed. *The Crockett Almanacks: Nashville Series, 1835-1838.* Chicago: The Caxton Club, 1955.
Melville, Herman. *The Confidence-Man: His Masquerade.* Edited by Hershel Parker. New York: W. W. Norton, 1971.
Miles, Guy S. "David Crockett Evolves, 1821-1824." *American Quarterly* 8 (1956): 53-60.
Mississippi Quarterly. 1948- .
Moseley, Elizabeth R. *Davy Crockett: Hero of the Wild Frontier.* Champaign, Ill.: Garrard, 1967.
Moving Picture World. April 23, 1910; May 7, 1910.
Murdock, Frank. *Davy Crockett; Or, Be Sure You're Right, Then Go Ahead.* In *Davy Crockett and Other Plays*, vol. 4 of *America's Lost Plays*, edited by Isaac Goldberg and Hubert Heffner. Princeton, N.J.: Princeton University Press, 1940. Reprint. Bloomington: Indiana University Press, 1963, pp. 115-48.
N. See Crockett, *Narrative.*
Parrington, Vernon Louis. "The Crockett Myth." In *Main Currents in American Thought*, vol. 2. New York: Harcourt, 1927, pp. 172-79.
Paulding, James Kirke. *The Lion of the West; Retitled The Kentuckian, or A Trip to New York: A Farce in Two Acts.* Revised by John Augustus Stone and William Bayle Bernard. Edited by James N. Tidwell. Stanford: Stanford University Press, 1954.
Peña, José Enrique de la. *With Santa Anna in Texas: A Personal Narrative of the Revolution.* Translated and edited by Carmen Perry. College Station: Texas A&M University Press, 1975.
People's Almanac. 1834-1842.
Porter, Kenneth W. "Davy Crockett and John Horse: A Possible Origin of the Coonskin Story." *American Literature* 15 (1943): 10-15.

Raspe, Rudolph Erich. *Singular Travels, Campaigns, and Adventures of Baron Munchausen.* Edited by John Carswell. New York: Dover Publications, 1960.
Readers' Guide to Periodical Literature. 1900- .
Robbins, J. Albert, et al. *American Literary Manuscripts: A Checklist of Holdings in Academic, Historical, and Public Libraries, Museums, and Authors' Homes in the United States.* 2nd ed. Athens: University of Georgia Press, 1977.
Rourke, Constance. *American Humor: A Study of the National Character.* New York: Harcourt, 1931.
_____. *Davy Crockett.* New York: Harcourt, 1934.
Sann, Paul. *Fads, Follies, and Delusions of the American People.* New York: Bonanza Books (Crown Publishers), 1967.
Schultz, Christian, Jr. *Travels on an Inland Voyage through the States of New-York, Pennsylvania, Virginia, Ohio, Kentucky and Tennessee, and through the Territories of Indiana, Louisiana, Mississippi and New-Orleans; Performed in the Years 1807 and 1808; Including a Tour of Nearly Six Thousand Miles.* Edited by Thomas D. Clark. Ridgewood, N.J.: Gregg Press, 1968.
Seelye, John. "The Well-Wrought Crockett: Or, How the Fakelorists Passed through the Credibility Gap and Discovered Kentucky." In *Toward a New American Literary History: Essays in Honor of Arlin Turner,* edited by Louis J. Budd et al. Durham, N.C.: Duke University Press, 1980, pp. 91-110.
Shackford, James Atkins. "The Author of David Crockett's Autobiography." *Boston Public Library Quarterly* 3 (1951): 294-304.
_____. "The Autobiography of David Crockett: An Annotated Edition." Ph.D. dissertation, Vanderbilt University, 1948.
_____. *David Crockett: The Man and the Legend.* Edited by John B. Shackford. Chapel Hill: University of North Carolina Press, 1956.
Shapiro, Irwin. *Yankee Thunder: The Legendary Life of Davy Crockett.* New York: Julian Messner, 1944.
Shapiro, Nat, ed. *Popular Music: An Annotated Index of American Popular Songs.* New York: Adrian Press, 1964. Vol. 1.
Simms, William Gilmore. *Michael Bonham: Or, The Fall of Bexar.* Richmond: John R. Thompson, 1852.
Slotkin, Richard. *Regeneration through Violence: The Mythology of the American Frontier, 1600-1860.* Middletown, Conn.: Wesleyan University Press, 1973.
Smith, David R. Letter (from the Disney Archives) to the author. February 10, 1981.
Smith, Seba. *The Life and Writings of Major Jack Downing, of Downingville, Away Down East in the State of Maine, Written by Himself.* Boston: Lilly, Wait, Colman, & Holden, 1833. Reprint. New York: AMS Press, 1973.
Smith, Richard Penn. *Col. Crockett's Exploits.* See "Crockett, David."
Southern Folklore Quarterly. 1937- .
The Spirit of the Times: A Chronicle of the Turf, Agriculture, Field Sports, Literature and the Stage. December 10, 1831-June 22, 1861. Weekly; title varies; publication and volume numbering not consistent before volume 6 (beginning February 20, 1836).
Stiffler, Stuart A. "Davy Crockett: The Genesis of Heroic Myth." *Tennessee Historical Quarterly* 16 (195'.): 134-40.

Tennessee Historical Quarterly. 1942- .
Thorpe, Thomas Bangs. "The Big Bear of Arkansas." In Cohen and Dillingham, eds. (see under), *Humor of the Old Southwest*, pp. 268-79.
Torrence, Robert M., and Whittenburg, Robert L. *Colonel "Davy" Crockett.* A genealogy. Washington, D.C.: Homer Fagan, 1956.
Walt Disney Productions. "The Ballad of Davy Crockett." Sheet music. Verses by Tom Blackburn; music by George Bruns. New York: Wonderland Music Co., 1954. See also Ford, Tennessee Ernie, and Hayes, Bill.
———. Initial television series: "Davy Crockett, Indian Fighter," 1954; "Davy Crockett Goes to Congress," 1955; "Davy Crockett at the Alamo," 1955. Follow-up television shows: "Davy Crockett's Keelboat Race," 1955; "Davy Crockett and the River Pirates," 1955. With Fess Parker and Buddy Ebsen. Bill Walsh, series producer.
———. *Davy Crockett, King of the Wild Frontier.* A film made from the first three television shows. With Fess Parker and Buddy Ebsen. Bill Walsh, producer. 1955; shortened, re-released, 1978.
———. *Davy Crockett and the River Pirates.* A film made from the fourth and fifth follow-up television shows. With Fess Parker and Buddy Ebsen. Bill Walsh, producer. 1956.
Wecter, Dixon. *The Hero in America: A Chronicle of Hero-Worship.* 1941. Reprint with introduction by Robert Penn Warren. New York: Charles Scribner's Sons, 1972.
Williams, Jerry T. *Southern Literature 1968-1975: A Checklist of Scholarship.* Boston: G. K. Hall, 1978.
Wright, Joseph, ed. *The English Dialect Dictionary.* London: Henry Frowde, 1898. Reprint. New York: Hacker Art Books, 1962. Vol. 2.
Yates, Norris B. *William T. Porter and the* Spirit of the Times: *A Study of the BIG BEAR School of Humor.* Baton Rouge: Louisiana State University Press, 1957.
Zanger, Jules. "The Frontiersman in Popular Fiction, 1820-60." In *The Frontier Re-examined*, edited by John Francis McDermott. Urbana: University of Illinois Press, 1967, pp. 141-53.

INDEX

Abbott, John S. C., 100, 136
ABC (American Broadcasting Company), 91, 95
Abingdon, Virginia, 11
Account of Colonel Crockett's Tour to the North and Down East (Whig political book), 46, 51-52, 99, 142-43
Adam: as example of myth, 55
Adams, Henry, 43
Adams, John Quincy, 43
Adventures of Davy Crockett, Told Mostly by Himself (Scribner's), 98
Adventures of Huckleberry Finn: Crockett compared to Tom Sawyer in, 21; "Little Davy" in bragging scene in, 77
Alabama River, 19
Alamo: battle of, 49-50, 93-95, 101-3, 145; Crockett's appearance and conduct in, 60-61; Crockett's death in, 6, 33, 50-54, 61; Crockett's supposed survival of, 60, 81
Alamo, The (United Artists), 60, 95-96, 113 (*illus.*), 145
Alamo museum, 62, 141
Albanese, Catherine L., 138
Alexander, Colonel Adam, 39-40
"Alex J. Dumas": pseudonym and hoax, 98-99

Allen, Charles Fletcher, 101, 135
Almanacs. *See* Crockett almanacs
American Antiquarian Society, 143
American Comic Almanac, 81
American Literature, 140
American Turf Register and Sporting Magazine, 87
Andrews, Ambrose: 110 (*illus.*)
"Another guess sort": Crockett idiom, 125
Apalachicola River, 27, 117
Apollo: Crockett compared to, 82
Appleseed, Johnny, 91
Arkansas River, 49
Arnold, William, 40-41
Arno Press, 143
Arpad, Joseph John, 71-73, 76-77, 80, 134-35, 137, 142-44
Atlanta, Georgia, 87
Augusta, Georgia, 87
Austin, Stephen F., 48
Autobiography: Crockett's. *See* Crockett, David; *Narrative of the Life of David Crockett*
Autobiography (Franklin), 5-6. *See also* Franklin, Benjamin
Autobiography of David Crockett (Garland, ed.). *See* Garland, Hamlin

Index

Backwoods life, xviii, 9-12, 17, 28, 80, 89-92; common sense and folk wisdom, 31-33, 56-57, 59, 80, 84-85; importance of humor, 56-57, 59, 62-67, 80, 84-85, 87-88, 97-98, 134-35; importance of hunting, 17-18, 35-36. *See also* Backwoodsman; Crockett, David; Tall tale; Tall talk

Backwoodsman: Crockett as image of, xix, 7, 20, 32-33, 36-37, 46, 59-60, 67-77, 88, 96, 100, 103; popular image of, xix, 20, 32-33, 36-37, 46, 56, 59-60, 63, 67-77, 103. *See also* Backwoods life; Crockett, David; Tall tale; Tall talk

Bad Day at Black Rock, 77
"Ballad of Davy Crockett" (Disney), 92-93, 144
"Ballad of Davy Crockett" (1834), 92, 144
Baltimore, Maryland, 13, 46
Bardstown, Kentucky, 80
Baton Rouge, Louisiana, 75-76
Bauer, Frederick E., 143
Beadle & Adams (publishers), 144
Beale, Charles T., 98-99
Bean's Creek, Tennessee, 18-19
Beard, Charles and Mary, 136
"Be always sure you're right—then go ahead": Crockett maxim, 6, 32, 45, 50, 66-67
Bear-Hunter, or, Davy Crockett as a Spy (Ellis), 102
Beasley, Major Daniel, 19-20
Beaty's Spring, Alabama, 116
Ben Hardin's Crockett Almanac: "A Sensible Varmint," 133 (*excerpt*). *See also* Crockett almanac
Beowulf: bragging in, 78; Grendel compared to bears, 35
Beverly Hillbillies, 93
Biddle, Nicholas, 70-71
"Big Bear of Arkansas" (Thorpe), 77, 87
Bird, Robert Montgomery, 102
Blackburn, Tom, 92
Black Warrior River, 22
Black Warrior's Town, Alabama, 22

Blair, Walter, 78, 90, 92-93, 100, 102, 136-37, 144-45
Blankenship, Russell, 136
Blount, Governor Willie, 25
Boone, Daniel, 74-75, 80, 91, 100, 135, 138, 145
Boone, Richard, 95
Boorstin, Daniel J., 138
Borgnine, Ernest, 77
Boston, Massachusetts, 81-82, 87-88
Botkin, Benjamin A., 92, 137, 144
Bowers, Jane, 144-45
Bowie, James, 95
Bragging. *See* Tall talk
Brown, Dee, 103-5, 137, 145
Bruns, George, 92
Buddha: as example of myth, 55
Buffalo Bill (William F. Cody), 135-36
Burgin, Adam, 49
Burnt Corn Creek, Alabama, 19
Burt, A. L. (publishers), 98
Bury My Heart at Wounded Knee (Brown), 103
Butler, William, 37-39

Canebrake: defined, 34
Captain Fear of Hell's Gulch: Crockett compared to, 78
"Captain Scott's Coon Story," 86-87, 131-32, 132-33 (*excerpt*)
Carey & Hart (publishers), 51-52, 83
Carey, Edward, 83-84
Carroll County, Tennessee, 34
Carson, Kit, 135-36
Caruthers, William Alexander, 71-72, 102
Castrillón, General Manuel Fernández, 53
Catron, Anna Grace, 135
Cawelti, John G., 139
CBS (Columbia Broadcasting System), 91
Chanticleer of the Wilderness Road (Le Sueur), 102
Chapman, John Gadsby: description of Crockett, 60-64, 141; paintings of Crockett, 61-62, 101, 107-8 (*illus.*), 141
Charles Scribner's Sons (publishers), 98

Charleston, South Carolina, 87
Cherokee Indians, xix, 9
Chickasaw Indians, xix, 26-27, 31
Chilton, Thomas: collaborator on Crockett's *Narrative*, 4-5, 15, 46, 97-98, 105, 142-43
Chittick, V.L.O., 137
Choctaw Indians, xix, 26-27
Choctawhatchee River, 27
Christ: as example of myth, 55
Cincinnati, Ohio, 87
Clark, Thomas D., 76-77
Clarke, Mathew St. Clair: author of first Crockett book, 3-4, 52, 83-85, 97-98, 105, 114-15, 124-25, 129-30, 142-43; motives for writing, 46, 70-74; occasionally more reliable than Crockett, 15, 65-66; plagiarist and hoaxer, 45-46, 67-77, 80. See also *Sketches and Eccentricities of Colonel David Crockett*
Clarksville, Texas, 49
Clay, Henry, 104
Clemens, Samuel, 10. See also Twain, Mark
Cocke, General John, 24
Cody, William F., 135
Coffee, Colonel John, 21-22
Col. Crockett, the Bear King (Ellis), 102
Col. Crockett's Exploits and Adventures in Texas (Smith), 51-53, 60, 88, 97-98, 101-4, 114-15, 125, 142-43; composition of, 51-52, 80, 83-85, 125; "A Useful Coonskin" (Crockett letter) in, 83-85, 88, 125, 126-29 (*excerpt*). See also Smith, Richard Penn; Garland, Hamlin
Collins, T. K. & P. G.: false name for Carey & Hart, 52
Colonel: as title, 31-32
Colonel Crockett: common name for David Crockett, 32
"Colonel Crockett and the Bear and the Swallows," 85-87
Colonel Crockett's Tour. See *Account of Colonel Crockett's Tour to the North and Down East*
Commedia dell'arte: bragging in, 78
Common sense: Crockett's as type of backwoodsman's, xxii, 8, 31-33, 41
Conecuh River, 27
Confidence-Man, The (Melville), 103
Conried, Hans, 94
Cooper, James Fenimore, xx, 88, 138
Cooper, Texas Jim, 141
Coosa River, 22-23
Courier and Enquirer, 53
Cravat, Nick, 94
Creek Indians, xix, 9, 18-28, 44, 134
Crockett: as common name for David Crockett, 32
Crockett, Ashley, 141
Crockett, Dame: in *Be Sure You're Right*, 89
Crockett, David: awareness of legend, xvii-xxiii, 4, 7, 33-34, 37, 44-47, 54, 58, 67, 70, 97-98, 105, 114-15, 125, 137; backwoods image, xix, 31-33, 41, 46, 56, 58-59, 63-64, 76-77, 107 (*illus.*), 110, 111; birth, 11; as campaigner, xviii, 8, 32, 35-41, 47-48, 84-85, 93-95, 104, 136, 139; childhood, 10-14; chronology, 146-50; common sense, 31-33, 41; courtship and marriage (Polly): 15, 21-28, (Elizabeth): 28-29, 65; death, 50-54, 60; as family man, 15, 17-18, 20, 28-29, 35-36, 39, 48-49; genealogy, condition of, 140; humor, sense of, 10, 37-41, 44-45, 56, 59-60, 62-67, 84-85, 94-95, 99, 104, 114-19, 116-33 (*excerpts*), 125, 129-32, 136-37; as hunter, 27-28, 33-35, 39, 61-62, 107 (*illus.*), 119-23 (*excerpts*); Indians, attitude toward, xvii-xxiii, 18-28, 44, 102, 117-19 (*excerpt*); Jackson, attitude toward, xvii-xxiii, 9-10, 25, 38, 44, 49-50, 66, 70-74, 134, 139; land bill, fight for, xxi, 33-37, 41-43; as legendary hero, xvii-xxiii, 50-60, 96-97, 104-5; as legislator, xvii-xxiii, 3-4, 33-39, 41-44, 62, 64, 93-95, 130, 136, 139; malaria, 30-31, 41; marksmanship, 14, 22, 35, 53, 60, 63, 67, 90-91, 105; monuments to, 11, 141-42; *Narrative*, composition of, style of, 3-8, 10, 25, 97-98, 116-25 (*excerpts*), 135; political purposes,

interest in presidency, xxii, 3-4, 8, 30, 46-47, 70-74; rifles of, as tool and symbol, 14, 35-36, 46, 141; as storyteller, 3-6, 8, 18, 22-23, 33-35, 41, 51-52, 58-59, 62-67, 83-88, 93-95, 97-99, 114-15, 116-33 (*excerpt*), 134-35, 142-43; style as a man, appearance, 50-54, 60-67, 90-91, 107-13 (*illus.*), 140-41. *See also* Crockett, Davy; *Davy Crockett* (titles); Democrats; Jackson, Andrew; *Narrative of the Life of David Crockett*; Whigs
Crockett, David (grandfather), 9
Crockett, Davy: beginnings in *Lion of West*, 88-96; as braggart, 78; as collective invention, 55-60; common name for Crockett, 32; as fictional character, 55-105; as godlike hero, 82-83; various images of, xvii, 59-60, 79, 83-92, 99-100; as persona in yarns, 85-86. *See also* Crockett, David; Wildfire, Nimrod; *Davy Crockett* (titles)
Crockett, Elizabeth Patton (second wife): character, 29, 31, 65; marriage to Crockett, 28-29, 35-36, 65; Texas homestead, 49
Crockett, James (uncle), 9
Crockett, John (father), 9-14, 29
Crockett, John (son), 33-34, 42-43
Crockett, Joseph (uncle), 9, 13
Crockett, Margaret (daughter), 28; (Mrs. Wiley Flowers), 48, 50
Crockett, Mary (Polly) Finley (first wife): Crockett's courtship of, 15-17; death, 28; marriage, 18-21, 26; in Disney series, 93-94; in *Wave High the Banner*, 104
Crockett, Rebecca Hawkins (mother), 10-11
Crockett, Rebeckah (daughter), 49
Crockett, Robert (son), 49
Crockett, William (uncle), 9
Crockett almanacs, 32, 59, 79-83, 85, 87-88, 104, 111 (*illus.*), 130, 136, 143-45; "Crockett's First Speech in Congress," 131 (*excerpt*); "A Sensible Varmint," 133 (*excerpt*)

Crockett's Exploits. *See Col. Crockett's Exploits and Adventures in Texas*
Crockett's Free-and-Easy Song Book, 144
"Crockett's March," 144
Cumberland Gap, 80
"Cut out": Crockett idiom, 116

Daily Louisville Public Advertiser, 68
Dame Crockett: in *Be Sure You're Right*, 89
Damon, Montgomery: in *Kentuckian in New York*, compared to Crockett, 71-72, 102
Daughters of the Texas Revolution: library in Alamo museum, 62, 108 (*illus.*)
David Crockett (Abbott), 100
David Crockett (Allen), 101, 135
David Crockett: The Man and the Legend. *See* Shackford, James A.
Davis, Curtis Carroll, 141
Davy Crockett. *See* Crockett, Davy; Crockett, David
Davy Crockett (Holbrook), 101, 137, 141
Davy Crockett (Rourke), xix, 79, 81-83, 136-37, 143-45
Davy Crockett (Selig Polyscope film), 145
Davy Crockett, American Comic Legend (Dorson), 79-80, 143-44
Davy Crockett and the River Pirates (Disney film), 92, 112 (*illus.*)
"Davy Crockett at the Alamo" (in Disney television series), 93-95
Davy Crockett—Frontier Hero (Blair), 102, 136
"Davy Crockett Goes to Congress" (in Disney television series), 93-95, 130
Davy Crockett: Hero of the Wild Frontier (Moseley), 102
"Davy Crockett: Indian Fighter" (in Disney television series), 93-95
Davy Crockett, Indian Scout (Republic Pictures), 145
Davy Crockett, King of the Wild Frontier (Disney film), 92-95
Davy Crockett; Or, Be Sure You're

Index   163

Right, Then Go Ahead (Murdock), 88-91, 144; seen by Hamlin Garland, 99-100; source for 1910 film, 145. See also Garland, Hamlin; Mayo, Frank; Murdock, Frank
Davy Crockett's Almanack: "Crockett's First Speech in Congress," 130-31 (*excerpt*). See also Crockett almanacs
Democrats: Crockett as loutish example of, 136; Crockett vs. Jacksonites, xvii-xxiii, 3, 43, 46, 70-74; Jacksonian politics, xviii-xx, 49-50, 82. See also Jackson, Andrew; Whigs
DeRose, Anthony Lewis, 140
DeVoto, Bernard, 9-10, 137
Dickinson, Almeron: survivor of the Alamo, 51
Dickinson, Samuel (publishers), 81
Dime novels: Crockett in, 101-2, 144; Indians in, xix, 102
Dimmitt, Richard B., 145
Disney, Walt: filmmaker and leader in television industry, 91. See also Walt Disney Productions
Disney Archives, 144
Disneyland (park), 91
Disneyland (television show), 91-95
Dobie, J. Frank: collection at University of Texas, 107 (*illus.*), 141
"Dog and the Raccoon": variants discussed, 83, 86-87, 131-33 (*excerpts*)
Doggett, Jim: in "Big Bear of Arkansas," 77, 87
Dorson, Richard M., 58, 79-80, 82, 84-86, 136-37, 143-44
Downing, Major Jack (Seba Smith), 143
Duck River, 17
"Dumas, Alex J.": pseudonym and hoax, 98-99

Earthquake: character in *Elkswatawa*, compared to Crockett, 72, 102
Earthquakes: of 1811-1812, in West Tennessee, 34
East Tennessee Historical Society's Publications, 135
Eaton, Senator John Henry, 104

Eaton, Peggy O'Neale Timberlake, 104
East Lynne (Wood), 89
Ebsen, Buddy, 93-95, 112 (*illus.*), 116
Education of Henry Adams, 43
Eleanor: heroine of *Davy Crockett; Or, Be Sure You're Right*, 89-91
Elder, Margaret, 14-15
Elk River, 17
Elkswatawa; Or, The Prophet of the West (French), 72, 135
Ellis, Edward, 101-2, 135, 144
Ellms, Charles, 81-82
Emuckfau Creek: battle of, 25
Enotachopco Creek: battle of, 25
Escambia River, 26-27, 117

Filmfacts, 145
Film Index, 145
Finch, Christopher, 93, 144
Finley, Mary. See Crockett, Mary (Polly) Finley
Finley, Mr. and Mrs.: Polly's parents, 15-17
Fink, Mike, 67
Finn, Huckleberry: compared to Crockett, 13-14
Fishwick, Marshall, 138
Fitzgerald, William, 44-45
Flowers, Margaret (Crockett) and Wiley, 28, 48, 50
Folklore: relationship to Crockett legend, 58-60, 79-88, 97, 104-5, 137, 142
Folmsbee, Stanley J., 5, 7, 99, 134-35, 141
Ford, Tennessee Ernie, 144
Fort Barrancas, 26
Fort Talladega, 23-24
Fort Mims, 19-20, 22, 29
Fort Montgomery, 27
Fort Strother, 22, 24, 28
Franklin, Benjamin: compared to Crockett, 5-6, 12, 143; Poor Richard compared to Crockett, 67; and history of almanacs, 79
Franklin County, Tennessee, 18-20, 28
Freeman, Mr.: in *Lion of the West*, 69-70
French, James Strange: *Elkswatawa*,

72, 102; and Clarke's *Sketches*, 3, 70-72, 142
Frontier life. *See* Backwoods life
Frontiersman. *See* Backwoodsman
Fulton, Arkansas, 49

Galbreath, Thomas, 11
Gale, Joseph, xviii
Gallows Dirk: backwoodsman in Irving's *History of New York*, 75-76
Galveston Bay, Texas, 53
Garland, Hamlin, 98-100, 135-36, 142, 144
Georgia Scenes (Longstreet), 51
"Georgia Theatrics" (Longstreet), 51
Georgie Russel: in Disney series, 94-95, 112 (*illus.*), 116. *See also* Ebsen, Buddy; Russell, George
Gibson, Major John, 21-22, 94, 116
"'Go Ahead' A March Dedicated to Colonel Crockett," 144
Goldberg, Isaac, 88-89, 144
Grant, James Edward, 95
Greene County, Tennessee, 11
Grendel: in *Beowulf*, compared to bears, 35

Hackett, James, 45-47, 61, 78, 105, 142; meets Crockett, 47; Wildfire's tall talk, 67-79, 110 (*illus.*)
"Half horse, half alligator": idiomatic phrase assigned to Crockett, 45, 67-79, 94, 130
Halley's comet: Crockett to wring its tail, 9-10, 83
Halliwell, Leslie, 145
Hammond's rangers, 23
Harding, Ben: figure in Crockett almanacs, 79-83; and Ben Hardin, Kentucky congressman, 80
Harper Brothers, 142
"Harricane": Crockett idiom, 34
Harris, Joel Chandler, 115
Harrison, Lowell, 137
Harrison, William Henry, 47
Harvey, Laurence, 95-96, 113 (*illus.*)
Hayes, Bill, 144
"Hazard, Harry": pseudonym, 102. *See also* Ellis, Edward

Heale, M. J., 138
Heffner, Hubert, 88-89, 144
Hercules: Crockett compared to, 82
Hickman County, Tennessee, 32
Hill, Hamlin, 78, 90, 92-93, 144
Hillabee faction (Creek nation), 24
History of New York (Irving), 75-76
Hit Parade, 92
Hofstadter, Richard, 139
Holbrook, Stewart, 101, 137, 141
Home Box Office, 93-95
Horseshoe Bend: battle of, 25
Horse-Shoe Robinson: character compared to Crockett, 103
Hough, Emerson, 100, 136
Houston, Sam, 49, 50, 95
How Did Davy Die? See Kilgore, Dan
"How the Man Came out of the Tree Stump": variant of Crockett yarn, 85-86
Hubbell, Jay, 136-37
Huckleberry Finn: compared to Crockett, 13-14. *See also Adventures of Huckleberry Finn*
Humor: in American periodicals and almanacs, 67-70, 79-83, 114; Crockett's, as example of popular American form, 8, 56-57, 99, 119, 135-37; in Crockett yarns, 83-88; as heroic and communal behavior, 37-38, 44-45, 59-67; in theater, 45-47, 67-79, 89-92
Hunnicut, Arthur, 145
"Hunters of Kentucky" (Woodworth), 74, 144
Huntsville, Alabama, 21-22

"I can whip my weight in wildcats": idiomatic phrase assigned to Crockett, 67-69, 73
Idioms: in Crockett's speech, described by Chapman, 62-67; in Crockett stories, 114-33. *See also* "Another guess sort"; "Be always sure you're right"; "Cut out"; "Half horse, half alligator"; "Harricane"; "I can whip my weight in wildcats"; "I know'd"; "In and about"; "I was a little wrathy"; "Kill more lickur";

"Root hog or die"; "Shot"; "Ticlur"; "Whapper of a lie"; "With the bark off"
"I know'd": Crockett idiom, 116; Crockett's use compared to usage in Wisconsin (Garland), 99
Iliad: bragging in, 78
"In and about": Crockett idiom, compared to usage in Wisconsin (Garland), 99
"In another guess sort": Crockett idiom, 125
Indians: assimilation into white communities, xix, 19, 44; in *Bury My Heart at Wounded Knee*, 103; control over Texas province, 48; Crockett in Creek War, 18-28, 44; in dime novels, xix, 102; in Disney film, 93; enmity among tribes, 26-27; as hunter-gatherers, 18; Indian removal and removal bill, xvii-xxiii, 20, 44, (in Disney film) 92-93; myth, 56; treaties broken by U.S., xx, 9; Trail of Tears, xviii, 19. *See also* Cherokee, Chickasaw, Choctaw, Creek, and Seminole Indians; Jackson, Andrew; Rourke, Constance
Indian removal bill. *See* Indians
Individualism. *See* Self-reliance
Inge, M. Thomas, 140
Irving, Washington, 75-76
"I was a little wrathy": Crockett idiom, 116

Jackson, Andrew, 4, 25, 30, 43-46, 66, 70-71, 134-39; commission to Crockett to wring comet's tail, 9-10, 83; in Creek War, xx, 21-26; Crockett's opposition to, xvii-xxiii, 25, 28, 44, 49-50, 66, 70-74, 134, 139; in Disney film, 93-94; and Indian removal bill, xvii-xxiii; in New Orleans, battle of, 74-75; in Pensacola, 26-27; in *Wave High the Banner*, 104. *See also* Democrats; Whigs
Jackson Gazette, xvii-xviii, xxii
Jackson, Tennessee, 37, 104
Jacksonites. *See* Democrats

Jacobs, Robert, 139
Jarvis, John Wesley, 76, 137
Jefferson County, Tennessee, 11
Jeremiah Johnson, 77
Jim Doggett: in "Big Bear of Arkansas," 77, 87
Job Snelling: in "A Useful Coonskin," 84-85, 126-29 (*excerpt*)
Joe: survivor of Alamo, 51
Johannsen, Albert, 144
Johnny Appleseed, 91
Journal of American Folklore, 140
Journal of American History, 140

Kennedy, John, 14
Kennedy, John Pendleton, 103
Kentuckian: as backwoods type, 67, 74-75
Kentuckian in New York (Caruthers), 71-72, 102
Kilgore, Dan, 52-54, 96-97, 100-101, 135, 145
"Kill more lickur": Crockett idiom, 73
King, Margaret, 137, 144
Kings Mountain: battle of, 10
Kingston Trio at Large, 145
Kitchen, Benjamin, 12
Knoxville, Tennessee, 11-12

Land bill, xix-xxii, 17, 33-38, 41-43
Lanson, Snooky, 92
LaSalandra, Cassandra, 141
"Lasalle, Charles E.": pseudonym, 102. *See also* Ellis, Edward
Last Command (Republic Pictures), 145
Lawrence County, Tennessee, 29, 31-32, 35
Leatherstocking: Cooper's character as backwoods type, compared to Crockett, xx, 88, 138
Legend: as concept and collective invention, xxiii, 18, 33-35, 104-5; Crockett's awareness of his, xvii-xxiii, 4, 7, 33-34, 37, 44-47, 54, 58, 67, 70, 97-98, 105, 114-15, 125, 137; defined, 55-60; opposed to fact, xix, 33-35, 50-54, 96-105, 142-43. *See also* Myth; Crockett, David; Crockett, Davy

Leithead, Edward, 144
Lemay, J. A. Leo, 137
Le Sueur, Meridel, 102, 137
Life and Adventures of Colonel David Crockett. See *Sketches and Eccentricities of Colonel David Crockett*; Clarke, Mathew St. Clair
Life and Times of Judge Roy Bean, 77
Life of David Crockett, the Original Humorist and Irrepressible Backwoodsman: as compilation of *Narrative, Colonel Crockett's Tour*, and *Crockett's Exploits*, 98-99, 125; unreliable except for *Narrative*, 98-99, 135-36, 142-43
Life of Martin Van Buren (Whig political book), 52, 142-43
Life on the Mississippi (Twain), 77
Lilly Library (Indiana University), 139, 144
Lime Stone River, 11
Lincoln County, Tennessee, 17
Lincolnton, North Carolina, 9
Link, Samuel, 136-37
Lion of the West (Paulding), 45-47, 58, 67-74, 88, 97, 142
Little Davy: in Twain's keelboat scene, 77
Little Rock, Arkansas, 49
Lofaro, Michael A., 137, 143-45
Longstreet, Augustus Baldwin, 51
Lovell, Coryell & Co. (publishers), 98
Ludlow, Noah, 74

McBride, Robert M., 141
Major Jack Downing (Seba Smith), 143
Malaria: frontier malady, 10; Crockett's, 30-31, 41
Manifest Destiny, 21, 56
Mark Twain, 10, 31, 60, 66, 76
Mark Twain's America (DeVoto), 9-10, 137
Matthews, Captain, 32
Mayer, Edwin Justus, 101, 137
Mayo, Frank, 88-91, 99-100. See also *Davy Crockett; Or, Be Sure You're Right*
Meadowcroft, Enid, 102

Meine Collection (University of Illinois), 143
Meine, Franklin J., 79, 82, 136-37, 143-45
Melodrama, 88-89
Melville, Herman, 103
Memphis, Tennessee, 40
Michael Bonham (Simms): character compared to Crockett, 102-3
Mike Fink, 67
Miles, Guy S., 137
Mississippi Quarterly, 140
Mississippi River, 34, 39-40, 81, 104
Mitchell, James, 65-66
Mobile: battle of, 19, 26
Montgomery Damon: in *Kentuckian in New York*, compared to Crockett, 71-72, 102
Montgomery, George, 145
Monuments: to Crockett, 11, 141-42
Morristown, Tennessee, 141
Moseley, Elizabeth, 102
Moving Picture World, 145
Munchausen, Baron: compared to Crockett, 78
Muppet Show, 77
Murdock, Frank, 88-91, 99-100, 144-45
Muscle Shoals, Alabama, 22
Myth: defined, 7, 55-60; as collective invention, 7, 17, 80, 105. See also Legend

Nacogdoches, Texas, 48
Narrative of the Life of David Crockett of the State of Tennessee (Crockett's autobiography): bear hunting in, 33-34, 39, 119-23; compared to Franklin's *Autobiography*, 4-6; composition and style of, 3-8, 10, 25, 97-98, 135; Crockett's politics in, 38, 46, 72; Crockett's yarn spinning, 57-58, 64-67, 134-35, 142-43; source for later works, 96-104, 114-25 (*excerpts*), 142-45
Nashville almanacs: probable hoax, 79-83, 97, 136. See also Crockett almanacs; Ellms, Charles; Seelye, John

Nashville, Tennessee: reaction to news of Crockett's death, 62
Natchez-under-the-Hill, 76
Navarro, Sánchez, 61
NBC (National Broadcasting Company), 91
New Orleans, 74, 87; battle of, 26, 74-75
New River, 13
New York City, 46, 81-82, 88
New York Historical Society, 140
Nick of the Woods (Bird), 102
Nietzsche, Friedrich, 57
Nimrod Wildfire: figure of Crockett in *Lion of the West*, 45-47, 58, 61, 67-74, 79, 110 (*illus.*), 142. *See also* Hackett, James; Paulding, James Kirke
North Carolina: land warrants in Tennessee, 36, 38

Obion River: Crockett's raft of staves, 39; Crockett homestead near, 33-35, 39, 65
Ohio River, 104
Old Farmer's Almanac, 79
Osgood, Samuel Stillman, 109 (*illus.*), 141

Paris, Tennessee, 44-45
Park Theater, 69
Parker, Fess, 88, 92-95, 96, 101, 112 (*illus.*), 129-31, 138-39, 145
Parrington, V. L., 7, 136
Patton, Elizabeth. *See* Crockett, Elizabeth Patton
Patton, George, 48-49
Patton, James, 28
Patton, William, 49
Patty Snaggs: in *Lion of the West*, 69
Paulding, James Kirke, 45-47, 58, 67-74, 88, 97, 105, 137, 142. *See also* Hackett, James; *Lion of the West*; Nimrod Wildfire
Peña, José Enrique de la, 52-54
Pensacola, Florida, 19, 26
People's Almanac, 81
Perry, Carmen, 52-54
Philadelphia, Pennsylvania, 46, 52, 87

Pirates Own Book (Ellms), 81
Polk, James K., 42, 104
Poor Richard's Almanack, 67, 79
Porter, Kenneth, 137
Porter, William T., 67-68, 87
Prometheus: as example of myth, 55; Crockett compared to, 82
Promised Land, 56

Raspe, Rudolph Erich, 78
Readers' Guide to Periodical Literature, 140
Red River, 49
Red Sticks: name for Creek warriors, 21-22, 44. *See also* Creek Indians
Reelfoot Lake, 34
Register of Debates in Congress, xvii-xviii
"Remember the Alamo" (Bowers), 144-45
Rifles: Crockett's, at shooting match, 14; as tool, 35-36; Philadelphia gift, 46; Old Betsy and Pretty Betsy, 141
Roaring Ralph: character in *Nick of the Woods*, compared to Crockett, 102
Robbins, J. Albert, 139
Rogersville, Tennessee, 9
"Root hog or die": Crockett idiom, 117
Rourke, Constance, xix, 79, 81-83, 102, 136-37, 143-45
Royal, Anne, 103-4
Russel, Georgie: based on George Russell in *Narrative*, 94, 116 (*excerpt*); character in Disney series, 94-95, 112 (*illus.*), 116. *See also* Ebsen, Buddy
Russell, Major William, 24, 26, 94
Ruysdael, Basil, 93

St. Louis, Missouri, 87
San Antonio, Texas, 49
San Augustine, Texas, 48-49
Sand Creek, battle of, 20
Sann, Paul, 138-39
Santa Anna, General Antonio López, 50-53
Sasquatch: as example of legend, 56

Sawyer, Tom: in *Huckleberry Finn*, Crockett compared to, 21
Schultz, Christian, Jr., 75-76
Scott, Captain Martin, 86-87
Scott, Sir Walter, 89
Scribner's (publishers), 98
Seaton, William, xviii
Seelye, John, 80-82, 104-5, 136-37, 143, 145
Self-reliance: subject of Crockett legend, xix, xxi, xxiii, 6, 12, 31-33, 36, 56. *See also* Backwoodsman
Selig Polyscope, 145
Seminole Indians, xviii, xix
"Sensible Varmint, A": 86-87, 131-32, 133 (*excerpt*)
Shackford, James A., xxii, 5, 8, 50, 52, 60-61, 80, 96-97, 100-101, 134-35, 139-43, 145
Shackford, John B., 52
Shapiro, Irwin, 101, 137
Shapiro, Nat, 144
Shoal Creek, 29, 31, 65
"Shot": Crockett idiom, 63
Siler, Jacob, 11-12
Simms, William Gilmore, 102-3
Sketches and Eccentricities of Colonel David Crockett (Clarke), 3-4, 15, 45-46, 52, 65-77, 80, 83-85, 97-98, 114-15, 125, 126, 129-30 (*excerpt*), 142-43. *See also* Clarke, Mathew St. Clair; *Narrative of the Life of David Crockett*
Slotkin, Richard, 138, 145
Smith, David R., 144
Smith, Richard Penn, 51-53, 60, 80, 83-85, 97-99, 101, 105, 114-15, 125, 142-43. See also *Col. Crockett's Exploits and Adventures in Texas*
Smith, Seba, 143
Snaggs, Patty: in *Lion of the West*, 69
Sockdolager! A Tale of Davy Crockett (Ellis), 144
Southern Folklore Quarterly, 140
Spirit of the Times, 47-48, 67, 70, 86-87, 130; "Captain Scott's Coon Story," 132-33 (*excerpt*)
Stanley, Helene, 94

Stiffler, Stuart A., 138
Story of Davy Crockett (Meadowcroft), 102
Story of the Wild West (Cody), 135
Sunrise in My Pocket (Mayer), 101

Tall tale: America as, 105; backwoods tradition of, 57, 117, 129; bears and hunters in, 18, 33-35; Crockett legend as, 18, 33-35, 57, 117, 129, 136-37
Tall talk: bragging in heroic traditions, 35, 67-83; in *Be Sure You're Right*, 89-92; Crockett's as example, 34, 57, 67-83, 115; in Disney series, 93-95; in *Narrative*, 97-98
Tallusahatchee, Alabama, 22-24
Tecumseh, 19
Tennessee Historical Quarterly, 140
Tennessee River, 21-22, 31
Texan Trailer; or, Davy Crockett's Last Bear Hunt (Ellis), 102
Thimblerig: in Disney series and *Crockett's Exploits*, 94
Thorpe, Thomas Bangs, 77, 87
"Ticlur" ("tickler"): Crockett idiom, 73
Tidwell, James N., 142
Tinkle, Lindsey, 49
Tracy, Spencer, 77
Trail of Tears, xviii. *See also* Indians
Travis, Colonel William, 51-53, 95-96
Tombigbee River, 19
Tom Sawyer: in *Huckleberry Finn*, Crockett compared to, 21
Torrence, Robert, 140
Turner & Fisher (publishers), 82
Twain, Mark, 10, 31, 60, 66, 76

Uncle Remus stories (Harris), 115
United Artists, 95, 105
"Useful Coonskin, A," 83-85, 114, 125-29 (*excerpt*), 137

Van Buren, Martin, 4, 47, 71. See also *Life of Martin Van Buren*
Van Doren, Carl, 101
Van Swearengen, Lieutenant: in "Cap-

tain Scott's Coon Story," 86-87
Villanueva, Candelaria, 61-62

"Wal, I'll be shot": Crockett idiom, 63
Walt Disney Productions: Davy Crockett television series, film, ballad, fad, xvii, 60, 79, 88, 91-96, 99, 105, 112 (*illus.*), 116, 129-31, 137-38, 144-45. *See also* "Ballad of Davy Crockett"; Disney, Walt; Ebsen, Buddy; King, Margaret; Parker, Fess
War of 1812: British fleet in Pensacola, 26; strategy against British and Indian alliance, 19
Washington, D.C., 40-44, 73, 93, 103
Washington Theater, 70, 105
Watauga Settlement, 9
Wave High the Banner (Brown), 103-4, 145
Wayne, John, 60, 95-96, 105, 145; legend compared to Crockett's, 96-97, 113 (*illus.*)
Weatherford, William, 19-20
Wecter, Dixon, 138
West Point, 22, 43
"Whapper of a lie": Crockett idiom, 31, 66
Whigs: and Crockett's candidacy, xxii, 3, 46-47, 70-74, 82; as promoters of Crockett books, 46, 51-52, 70-74, 142. *See also* Democrats
White, General James, 24
Whittenburg, Robert, 140
Widmark, Richard, 95
Wildfire, Nimrod: figure of Crockett in *Lion of the West*, 45-47, 58, 61, 67-74, 79, 142
Williams, Colonel John, 38
Williams, Jerry T., 140
Wilson, Abraham, 14
Winchester, Marcus, 40
Winchester, Tennessee, 19, 21
"With the bark off": Crockett idiom, 129
Wonderful World of Disney, 91
Wood, Ellen, 89
Woodworth, Samuel, 74, 144
Wounded Knee: battle of, 20, 103

Yankee Thunder (Shapiro), 101
Yarn spinning: customs of, 8, 22-23, 59-65, 85, 115, 119, 129; and folk tales, 83-88; narrator in, 85-86; in print media, 69, 87-88, 104-5. *See also* Crockett, David; Tall tale; Tall talk
"Young Lochinvar" (Scott): character in, compared to Crockett, 89

Zanger, Jules, 139

About the Author

RICHARD BOYD HAUCK is Abe Levin Professor of Humanities at the University of West Florida in Pensacola. He is the author of *A Cheerful Nihilism: Confidence and "The Absurd" in American Humorous Fiction*.